William Guy Peck

Complete Arithmetic

Theoretical and Practical

William Guy Peck

Complete Arithmetic
Theoretical and Practical

ISBN/EAN: 9783337400835

Printed in Europe, USA, Canada, Australia, Japan

Cover: Foto ©berggeist007 / pixelio.de

More available books at **www.hansebooks.com**

Davies & Peck's United Course.

Complete Arithmetic

THEORETICAL AND PRACTICAL.

BY

William G. Peck, Ph. D., LL. D.,

*Professor of Mathematics and Astronomy in Columbia College,
and of Mechanics in the School of Mines.*

A. S. BARNES & COMPANY.

NEW YORK, CHICAGO AND NEW ORLEANS.

PUBLISHERS.

PUBLISHERS' NOTICE.

PECK'S MATHEMATICAL SERIES.

CONCISE, CONSECUTIVE, AND COMPLETE.

NOTE.—*Teachers and others discovering errors in any of the above works will confer a favor by communicating them to us.*

PREFACE.

THE object of the following work is to present, in logical order and within moderate limits, all the fundamental principles of arithmetic, together with their most important applications to the wants of the student, the artisan, and the man of business.

It commences with the simplest elements, and progresses by natural steps to the highest and most complex operations. All superfluous matter has been omitted, and great care has been exercised to avoid needless multiplicity of cases and rules; but in no instance has any essential principle been omitted or unnecessarily abbreviated.

It is believed that the definitions are plain and concise; that the principles are stated clearly and accurately; that the demonstrations are full and complete; that the rules are perspicuous and comprehensive; and finally, that every branch of the subject is amply illustrated by well-graded examples and problems.

The order of logical development is thought to be simple and practical. the method of treating successive subjects being uniform and essentially as follows: 1°. All necessary definitions are given; 2°. A few mental exercises are then introduced, being so worded and so arranged as to lead the pupil to a knowledge of the fundamental principles of the subject under consideration; 3°. The principles thus developed are used in demonstrating the required rule; and 4°, The rule is then illustrated and enforced by a sufficient number of graded examples and problems.

The Author takes this opportunity to thank the many teachers who have aided him by valuable suggestions and criticisms.

COLUMBIA COLLEGE,
  Sept. 14, 1877.

CONTENTS.

NOTATION AND NUMERATION.

1. A **Unit** is a single thing; as, *one pound, one foot, one day.*

2. A **Number** is a unit, or a collection of units; as, *one pound, three days, five feet.*

3. Arithmetic is the science of numbers.

It treats of the properties and relations of numbers, and of the methods of computation by means of numbers.

NOTE.—In what follows, the expressions 1°, 2°, 3°, etc., are read *first, second, third*, etc.

FORMATION OF NUMBERS.

4.—1°. Numbers from *one* to *ten* are formed by collecting simple units, or *ones.*

A single unit is called **one**; one and one more are **two**; two and one more are **three**; three and one more are **four**; and so on, to **ten**.

2°. Numbers between *ten* and *one hundred* are formed by collecting *tens* and *ones.*

One *ten* and *one* are **eleven**; one *ten* and *two* are **twelve**; one *ten* and *three* are **thirteen**; and so on to *two tens*, or **twenty**. Two *tens* and *one* are **twenty-one**; two *tens* and *two* are **twenty-two**; and so on to *three* *tens* or *thirty*. Four *tens* are **forty** ; five *tens* are **fifty** ; and so on to *ten-tens*, or **one hundred**. The intermediate numbers between thirty and forty, forty and fifty, and so on, are formed in the same manner as those between twenty and thirty.

3°. Numbers between *one hundred* and *one thousand* are formed by collecting *hundreds, tens,* and *ones;* numbers between *one thousand* and *ten thousand* are formed by collecting *thousands, hundreds, tens,* and *ones;* and so on, indefinitely.

Numbers formed by collecting *ones,* in the manner just explained, are called **integers ;** they are also called integral, or whole numbers.

CLASSIFICATION OF NUMBERS.

5. Numbers are divided into two classes, *abstract* and *denominate.*

An **Abstract Number** is a number whose unit is not named; as, *five, seven, eleven.*

A **Denominate Number** is a number whose unit is named; as, *three pounds, six miles, seven months.*

Denominate numbers are sometimes called *concrete* numbers.

A denominate number may be either *simple* or *compound.*

It is a **Simple Number** when all the units of the collection are of the same *name* or *denomination ;* as, *eight yards, eleven ounces, five feet.*

It is a **Compound Number** when all the units of the collection are not of the same denomination ; as, *three feet and six inches, four hours and twenty minutes, two pounds and eleven ounces.*

Note.—All *integers* are simple numbers.

NOTATION AND NUMERATION.

6. Notation is the method of writing numbers by means of figures, or of letters.

Numeration is the method of reading written numbers.

FIGURES.

7. The following figures are used in the **common,** or **decimal system of notation**:

0,	1,	2,	3,	4,	5,	6,	7,	8,	9.
naught,	one,	two,	three,	four,	five,	six,	seven,	eight,	nine.

These figures, taken separately, are called **digits.** The first one, named **naught,** is also called a **cipher,** or **zero**; it stands for **no number.** The remaining ones are called **significant figures**; they stand for the numbers written below them.

Figures are not numbers, but it will often be convenient to speak of them as such; in these cases, it is to be understood that we refer to the numbers which the figures represent.

PLACES OF FIGURES.

8. If several figures are written in a line, the one on the right is said to stand in the **first place,** the one next to the right stands in the **second place,** the one next to it in the **third place,** and so on. Thus, in the expression 3784, 4 stands in the *first place,* 8 in the *second place,* 7 in the *third place,* and 3 in the *fourth place.*

ORDERS OF UNITS.

9. The number **one** is called **a unit of the first order**; the number **ten,** regarded as a collection of *ones,* is called **a unit of the second order**; **one hundred,** regarded as a collection of *tens,* is called **a unit of the third order**;

and so on indefinitely, the unit of each succeeding order being *ten times* that of the next lower one.

The unit of the fourth order is **one thousand**; the unit of the fifth order is **ten thousand**; the unit of the sixth order is **one hundred thousand**; the unit of the seventh order is **one million**; and so on.

Units whose order is not named are supposed to be units of the first order, or *ones*.

GENERAL PRINCIPLES OF NOTATION AND NUMERATION.

10. Numbers are written and read in accordance with the following principles:

1°. *The same digit always represents the same number of units.*

2°. *The order of units represented is denoted by the place in which the digit stands.*

3°. *A cipher standing in any place shows that the number contains no units of that order.*

Thus, the expression 777 denotes 7 *hundreds,* 7 *tens,* and 7 *units,* that is, it stands for *seven hundred and seventy-seven.* In like manner, the expression 507 denotes 5 *hundreds,* 0 *tens,* and 7 *units,* that is, it stands for *five hundred and seven.* The expression 240 stands for *two hundred and forty.*

NOTE.—Places of figures and orders of units are counted from right to left, but numbers are written and read from left to right.

EXAMPLES IN NOTATION AND NUMERATION.

11. Any number less than one thousand may be written by the following

.   R U L E .

Begin at the left and write the figures that denote the hundreds, tens, and units, in their proper order.

EXAMPLES.

Write the following numbers :

1. Fifty seven.   *Ans.* 57.
2. Ninety four.   *Ans.* 94.
3. One hundred and sixty nine.   *Ans.* 169.
4. Nine hundred and fourteen.   *Ans.* 914.

5. Three hundred and sixty.   15. Two hundred and nine.
6. Two hundred and seven.   16. Five hundred and fifty.
7. Nine hundred and eight.   17. Six hundred and nine.
8. Seven hundred and seven.   18. Ninety seven.
9. Nine hundred and ninety.   19. One hundred and ten.
10. One hundred and twelve.   20. Two hundred and six.
11. Three hundred and four.   21. Six hundred and sixty.
12. Eight hundred and sixty.   22. Four hundred and five.
13. Eight hundred and four.   23. Seven hundred and six.
14. Three hundred and eight.   24. Six hundred and seven.

Any written number less than one thousand may be read by the following

RULE.

Begin at the left and read the hundreds, tens, and units, in their order, translating figures into words.

EXAMPLES.

Read the following numbers:

1. 29.   *Ans.* Twenty nine.
2. 107.   *Ans.* One hundred and seven.
3. 118.   *Ans.* One hundred and eighteen.

4. 506.   8. 270.   12. 186.   16. 999.
5. 670.   9. 809.   13. 204.   17. 400.
6. 977.   10. 422.   14. 309.   18. 207.
7. 835.   11. 109.   15. 470.   19. 554.

NOTE.—Before reading a number, let the pupil name the unit of each order of figures, beginning at the right ; thus, *units, tens, hundreds.*

From what precedes, we see that *notation* is the operation of translating numbers from *words* into *figures,* and that *numeration* is the operation of translating numbers from *figures* into *words.*

The digits of a number indicate natural parts into which it may be separated. Thus, 986 may be separated into the three parts, 900, 80, and 6, each of which has its own unit.

PERIODS OF FIGURES.

12. Written numbers, containing more than three figures, are separated into **periods** of three figures each, beginning at the right hand; the left-hand period may contain less than three figures.

The first period, counting from the right, is called the **period of units**; the second is called the **period of thousands**; the third is called the **period of millions**; and so on, as shown in the following table, called

THE NUMERATION TABLE.

Units of periods.	Trillions,			Billions,			Millions,			Thousands,			Units.		
	hundreds of	tens of	units of	hundreds of	tens of	units of	hundreds of	tens of	units of	hundreds of	tens of	units of	hundreds	tens	units
Periods. . .	3	1	7,	8	3	2,	4	1	5,	8	1	6,	7	8	3

The number written above is read, 317 *trillions,* 832 *billions,* 415 *millions,* 816 *thousands,* 783.

The unit of the first period is the simple unit **one** ; the unit of the second period is **one thousand** ; the unit of the third period is a thousand times one thousand, or **one million** ; and so on, as indicated in the table, the unit of each period being equal to *a thousand times* that of the next lower one.

The table may be continued to any desired extent; the units of the next succeeding periods are **quadrillions, quintillions, sextillions, septillions, octillions,** etc.

Every period, except the left-hand one, must contain three figures, but they may all be ciphers. Periods that contain three figures are said to be *complete*.

Periods are written and read as explained in the last article. In writing, we make them all complete, *except* the one on the left; in reading, we name the unit of each, *except* the one on the right.

ADDITIONAL EXAMPLES.

13. Any number whatever may be written by the following

RULE.

Begin at the left and write each period in order, separating it from the following one by a comma.

Write the following numbers:

1. Ten thousand, two hundred and six. *Ans.* 10,206.

2. One hundred and fourteen thousand, eight hundred and seventy nine. *Ans.* 114,879.

3. Seven hundred and fifty thousand, three hundred and eighty nine. *Ans.* 750,389.

4. Nine hundred thousand, three hundred and fifty.

5. Six million, one hundred and sixty nine thousand, four hundred and thirty seven.

6. Seventy six million, four hundred thousand, one hundred.

7. 22 billion, 103 million, 576 thousand, 102.

8. 102 trillion, 125 million, 403.

9. 8 trillion, 7 billion, and 76.

10. 41 quadrillion, 817 trillion, 217 billion.

11. 107 quintillion, 200 million, 757 thousand, 365.

12. 14 billion, 74 million, 231 thousand, and 5.

Any written number may be read by the following

RULE.

I. Begin at the right and point it off into periods of three figures each; the left-hand period may contain less than three figures.

II. Begin at the left and read the periods in their order, naming the unit of each, except that on the right.

NOTE.—After the number is pointed off, the pupil should name each period, beginning at the right; thus, *units, thousands, millions, billions,* etc.

EXAMPLES.

Read the following numbers:

1. 104217. *Ans.* One hundred and four thousand, two hundred and seventeen.

2. 2304516. *Ans.* Two million, three hundred and four thousand, five hundred and sixteen.

3. 1001010. *Ans.* One million, one thousand, and ten.

4. 825314715. *Ans.* 825 million, 314 thousand, 715.

5. 7416.	13. 6003021715.
6. 23562.	14. 4785003298.
7. 475437.	15. 12303492816.
8. 284871.	16. 117723326419.
9. 1284576.	17. 8843412956.
10. 4534218.	18. 543521798612.
11. 88334172.	19. 254321496.
12. 24137652.	20. 1546973200849.

ROMAN NOTATION.

14. Roman Notation is the method of expressing numbers by letters. The letters used and the values they express are shown below:

Letters . . . I, V, X, L, C, D, M.

Values, . . . 1, 5, 10, 50, 100, 500, 1000.

Other numbers are expressed by combining these letters according to the following principles:

1°. *If a letter is repeated, the number that it denotes is repeated.*

2°. *If a letter that denotes a less number is written after one that denotes a greater number, the value of the latter is increased by that of the former.*

3°. *If a letter that denotes a less number is written before one that denotes a greater number, the value of the latter is diminished by that of the former.*

If a letter that denotes a less number is written between two that denote greater numbers, it diminishes the latter, but does not affect the former.

The method of applying these principles is shown in the following

TABLE.

I denotes 1	XI denotes 11	XXX denotes 30	CCCC denotes 400
II " 2	XII " 12	XL " 40	D " 500
III " 3	XIII " 13	L " 50	DC " 600
IV " 4	XIV " 14	LX " 60	DCC " 700
V " 5	XV " 15	LXX " 70	DCCC " 800
VI " 6	XVI " 16	LXXX " 80	DCCCC " 900
VII " 7	XVII " 17	XC " 90	M " 1000
VIII " 8	XVIII " 18	C " 100	MM " 2000
IX " 9	XIX " 19	CC " 200	MDCCCLXXV de-
X " 10	XX " 20	CCC " 300	notes 1875.

EXAMPLES.

Read the following numbers:

1. XXXIX.
2. XCLVIII.
3. MDCXIX.
4. DCCLIX.

5. MMDXXXII.
6. DCCXLIII.
7. DCCCCXC.
8. CCCLXXXIII.

Write the following numbers by the Roman method:

9. 42.
10. 84.
11. 119.
12. 1,214.

13. 2,940.
14. 3,317.
15. 2,150.
16. 1,555.

NOTE.—A dash over a number written in Roman numerals increases the number 1,000 times. Thus, $\overline{\text{XXX}}$ stands for 30,000.

REVIEW QUESTIONS.

(**1.**) What is a unit? Example? (**2.**) What is a number? Example? (**3.**) What is arithmetic? What does it treat of? (**4.**) Explain the formation of numbers from *one* to *ten;* from *ten* to *one hundred,* etc. What is an integer, or whole number? (**5.**) How are numbers classified? What is an abstract number? A denominate number? When is a denominate number simple? When compound? (**6.**) Define notation. Numeration. (**7.**) Name the ten digits. What other names has the figure *naught?* Which are significant figures? (**8.**) Explain what is meant by the place of a figure? (**9.**) Explain what is meant by orders of units, and give the names of the orders up to the seventh. (**10.**) What are the three principles of decimal notation? How are places and orders counted, and how are numbers written and read? (**11.**) Give the rules for writing and reading any number less than 1000. (**12.**) What are periods of figures? Name the units of the first six periods. (**13.**) Give the general rules for notation and numeration. (**14.**) What is Roman notation? Explain the method of writing numbers in this system.

FUNDAMENTAL OPERATIONS.

I. ADDITION.

DEFINITIONS.

15. **Addition** is the operation of finding the sum of two or more numbers.

16. The **Sum** of two or more numbers is a number that contains as many units as the given numbers taken together. Thus, 5 *days* is the sum of 3 *days* and 2 *days*.

The numbers added, and their sum, must be *similar*.

17. **Similar Numbers** are those that have the same unit. Thus, 3 *yards* and 7 *yards* are similar, but 3 *yards* and 7 *days* are not similar.

Abstract numbers are always similar, because they have the same unit.

MENTAL EXERCISES.

1. John has 4 *apples* and James has 5 *apples;* how many have both? How many *apples* are 4 *apples* and 5 *apples?* How many are 4 and 5? 5 and 4?

2. Frank had 8 *marbles* and Peter gave him 7 more; how many had he then? What is the sum of 8 *marbles* and 7 *marbles?* What is the sum of 8 and 7? of 7 and 8?

3. An arithmetic class consists of 5 *boys* and 7 *girls;*

how many *pupils* are there in the class ? How many are
5 and 7 ? 7 and 5 ?

EXPLANATION.—Because both *boys* and *girls* are *pupils*, the numbers to be added are similar, although they appear to be dissimilar.

4. What is the sum of 5 *dollars,* 6 *dollars,* and 9 *dollars?*
What is the sum of 5, 6, and 9 ? of 6, 9, and 5 ? of 9, 6,
and 5 ?

5. A farmer has 6 *oxen,* 9 *cows,* and 8 *calves;* how many
cattle has he ? How many are 6, 9, and 8 ?

6. What is the sum of 10, 12, 8, and 4 ? of 4, 8, 10, and
12 ?

EXPLANATION OF SIGNS.

18. The **sign of addition,** +, is called **plus ;** when
placed between two numbers, it shows that the second is
to be added to the first. Thus, the expression 4 + 5 shows
that 5 is to be added to 4.

The **sign of equality,** =, indicates that the expressions between which it is placed are equal to each other.
Thus, 4 + 5 = 9, indicates that the sum of 4 and 5 is
equal to 9.

An expression of equality between numbers is called an
Equation ; the part on the left of the sign of equality is
the **first member** and the part on the right is the **second
member.** Thus, in the equation 9 + 8 = 17 the part
9 + 8 is the *first member* and the part 17 is the *second
member.*

MENTAL EXERCISES.

1. 4 + 8 + 7 = how many?

NOTE.—Let the pupil supply the second member and then read
the equation.

2. $4 + 5 + 7 + 2 + 1 =$ how many?

3. 7 *men* $+ 4$ *men* $+ 3$ *men* $+ 9$ *men* $=$ how many *men?*

4. 3 *dollars* $+ 2$ *dollars* $+ 9$ *dollars* $=$ how many *dollars?*

5. 6 *oxen* $+ 12$ *cows* $+ 7$ *calves* $=$ how many *cattle?*

6. $6 + 9 + 4 + 3 + 9 + 2 + 1 =$ how many?

7. $4 + 9 + 5 + 6 + 3 + 4 + 8 = ?$

8. The sum of 4, 9, 3, 2, 7, 6, and 5 *equals* how many?

9. 4 *yards* $+ 9$ *yards* $+ 11$ *yards* $+ 8$ *yards* $= ?$

NOTE.—The signs of interrogation in examples 7 and 9 indicate that the second members are to be supplied by the pupil.

10. $14 ft. + 3 ft. + 7 ft. + 9 ft. + 10 ft. = ?$

11. $9 + 9 + 7 + 3 + 6 + 8 + 7 + 4 = ?$

Let the pupil add the following columns:

(12.)	(13.)	(14.)	(15.)	(16.)	(17.)	(18.)	(19.)	(20.)	(21.)
6	7	3	9	5	4	7	6	7	3
7	4	8	2	1	2	8	4	2	6
8	3	2	8	8	2	7	4	4	7
5	9	8	3	2	4	6	5	3	3
4	8	3	9	9	4	2	5	5	4
2	4	5	6	3	5	3	7	7	9
3	7	4	5	7	5	4	8	2	1
8	5	9	4	6	6	1	9	9	8
1	2	6	7	4	9	5	2	6	2

Sum

NOTE.—The operation of adding a column of figures should be abbreviated by simply naming the result of each step. Thus, in example 12, the pupil should say 1, 9, 12, 14, 18, 23, 31, 38, 44.

The exercise may be varied by adding each column from top to bottom; also by adding the lines horizontally both forward and backward.

19. The operation of addition depends on the following principles:

1°. *Any number is equal to the sum of all its parts.*

2°. *The sum of two or more numbers is equal to the sum of all their parts.*

20. Let it be required to find the sum of 564, 783, and 688.

EXPLANATION.—Having written the numbers so that units of the same order stand in the same column, we begin at the right and add each column separately. The sum of 8, 3, and 4, is 15 *units*, that is, 1 *ten* and 5 *units;* we write the 5 in the column of units and *carry forward* the 1 and add it to the column of tens. The sum of the tens, thus increased, $1 + 8 + 8 + 6$, is 23 *tens*, that is, 2 *hundreds* and 3 *tens;* we write the 3 in the column of tens and *carry forward* the 2 and add it to the column of hundreds. The sum of the hundreds, thus increased, $2 + 6 + 7 + 5$, is 20 *hundreds*, that is, 2 *thousands*, and 0 *hundreds;* as this is the last column, we set down the entire sum. The resulting number, 2,035, is the required sum, because it is the sum of all the *units, tens*, and *hundreds* of the given numbers (Art. **19**).

OPERATION.

```
 564
 783
 688
────
2035
```

In like manner other numbers may be added; hence, we have the following

RULE.

I. Write the numbers so that units of the same order shall stand in the same column.

II. Add the column of units; set down the simple units of the sum, and if there are any tens, carry them forward and add them to the next column,

III. Add the column of tens; set down the simple

tens, and if there are any hundreds, carry them forward and add them to the next column.

IV. Continue this operation till all the columns have been added. Set down the entire sum of the last column.

EXAMPLES.

Perform the following additions:

(1.)	(2.)	(3.)	(4.)
315	29	215	8261
423	814	27	3042
719	302	891	171
Sum, 1457	1145	1133	11474

The rule holds good for all simple numbers, whether abstract or denominate.

(5.)	(6.)	(7.)	(8.)
451 *feet.*	365 *days.*	187 *pounds.*	124 *things.*
817 *feet.*	821 *days.*	203 *pounds.*	287 *things.*
302 *feet.*	900 *days.*	866 *pounds.*	59 *things.*
917 *feet.*	76 *days.*	771 *pounds.*	803 *things.*
2487 *feet.*	2162 *days.*	2027 *pounds.*	1273 *things.*

PROOF OF ADDITION.—*Perform the operation by commencing at the top of each column, and adding downward. The sum should be the same as before.*

NOTE.—Every operation in addition should be proved.

(9.)	(10.)	(11.)	(12.)	(13.)
9,102	8,760	25,678	87 *feet.*	62,743
479	325	3,002	236 *feet.*	4,321
73	512	21,001	1,443 *feet.*	78,731
810	786	715	2,010 *feet.*	1,239
4,312	1,420	1,630	7,818 *feet.*	4,241

(14.)	(15.)	(16.)	(17.)	(18.)
27 *yards.*	7,478 *days.*	117,064	2,571	2,476
135 *yards.*	423 *days.*	92,973	1,701	7,884
7,271 *yards.*	79 *days.*	827,569	973	3,349
185 *yards.*	8,102 *days.*	1,351	2,045	5,876

19. Add 7,384; 326; 6,780; and 57.　*Ans.* 14,547.

20. Add 6,740; 9,745; 5,769; 8,031; 6,543; 2,052; and 9,999.　　　　　　　　　　　　*Ans.* 48,879.

21. Add 89; 4,500; 423; 2,024; 5,408; 60,546; 9,401.

22. Add 83,746 *yards ;* 2,478 *yards ;* 692,577 *yards ;* 456 *yards ;* and 7 *yards.*

23. Add 935,473 *dollars ;* 262 *dollars ;* 13,897 *dollars ;* 598,453 *dollars ;* 25 *dollars ;* 3,734 *dollars ;* and 72,405 *dollars.*

The sign $ written before a number signifies *dollars ;* thus, the expression $120 is read 120 *dollars.*

24. Find the sum of $93,180; $279; $8,711; $371,800; $65 ; and $212,818.

25. Add 3,415; 17,382; 81,845; 162,345; and 8,342.

26. Add 8,492 *feet ;* 14,592 *feet ;* 112,897 *feet ;* and 117,712 *feet.*

27. Add $8,842; $31,887; $113,214; and $887,319.

28. Add 385,842; 112,817; 32,413; and 33,335.

29. Add $88,141; $32,314; $141,003; and $89,947.

30. Add 114,312; 87,808; 3,214; 896; and 87.

31. Add 8,730; 3,021; 785; 879; and 92.

32. Add $87; $78; $114; $289; $176; and $95.

33. 42,314 *yds. ;* 119,342 *yds. ;* 8,962 *yds. ;* 8,962 *yds.*

34. Add 17,439; 410,864; 842,317; 345,876; 79,884; and 18,719.

35. Add 714,312; 182,416; 312,867; 382,843; 79,816; and 43,115.

36. Find the sum of 3,345,816; 2,882,314; 387,892; 4,381,500; 2,874,316; and 887,342.

37. Add 188,841; 362,817; 411,217; 336,425; 814,316; and 45,554.

38. Add 214,333; 286,329; 851,426; 303,249; 12,456; 17,324; and 22,404.

39. Add 3,329,941; 187,693; 821,436; 2,227,438; 132,314; and 283,304.

40. Add 193,391; 4,180,280; 7,814,312; 88,430; 92,872; and 64,428.

41. Add 112,847; 186,320; 662,641; 3,400,300; 2,810,000; and 749,209.

42. Add 682,817; 336,336; 4,150,209; 2,390,374; and 86,810,304.

Dollars and cents may be written together, the cents being separated from the dollars by a point. Thus, the expression $17.84 is read 17 dollars and 84 cents.

Dollars and cents may be added like simple numbers. In writing them down, the separating points must stand in the same column.

(43.)	(44.)	(45.)	(46.)	(47.)
$18.73	$5.83	$186.40	$413.30	$2,234.75
23.47	10.19	75.75	325.15	3,821.62
15.62	27.03	37.18	414.82	911.94
7.91	11.94	201.92	97.45	89.69
112.13	203.07	184.42	111.32	10,312.41
648.21	211.46	36.35	202.16	9,102.70
73.19	305.24	41.15	113.27	25,444.33
19.06	802.41	72.27	814.42	42,829.77
35.62	111.37	94.79	316.81	11,312.48

48. What is the sum of $8,311.25, $27,494.62, $143,596.22, $155,463.79, $292,986.48, $382,811.67, $482,884.20, and $919,902.20 ?

49. Add $2,863,747.25, $3,894,511.82, $8,818,416.20, $215,714,381.46, $747,719.87, and $59,107,411.28.

50. Find the sum of $53.42, $881,16, $416.49, $1,381.40, $88.88, $210.29, $6.49, $511.11, $16.84, and $2,256.00.

PRACTICAL PROBLEMS.

21. A **Problem** is a question proposed that requires an answer. The operation of finding the answer is called the **solution of the problem.**

Solve the following problems :

1. A farmer sold a span of horses for $318, two pairs of oxen for $420, and six cows for $290 ; how much did he receive ?   *Ans.* $1,028.

2. A man bought a house for $24,500, paid $1,675 for repairs, $3,140 for furniture, $375 for taxes, and then sold the whole for $2,155 more than the cost ; what did he receive ?   *Ans.* $31,845.

ABBREVIATIONS.— In the following problems, *lbs.* stands for *pounds ; ft.* for *feet ; yds.* for *yards ;* and *bu.* for *bushels.* Other abbreviations will be explained in their proper places.

3. A wagon is loaded with 5 boxes of goods ; the first weighs 473 *lbs.*, the second 392 *lbs.*, the third 479 *lbs.*, the fourth 1,217 *lbs.*, and the fifth 376 *lbs.* ; what is the weight of the entire load ?   *Ans.* 2,937 *lbs.*

4. The first car of a freight train contains 8,117 *lbs.* the second 11,819 *lbs.*, the third 9,156 *lbs.*, the fourth 8,884 *lbs.*. the fifth 10,398 *lbs.*, and the sixth 9,982 *lbs.* ; how many pounds are there in all ?   *Ans.* 58,356 *lbs.*

5. A farm contains 79 *acres* of woodland, 63 of pasture land, 50 of meadow land, and 73 of arable land; how many acres in the farm? *Ans.* 265 *acres.*

6. A factory turned out 702 *yds.* of cloth on Monday, 1,023 *yds.* on Tuesday, 1,107 *yds.* on Wednesday, 997 *yds.* on Thursday, 910 *yds.* on Friday, and 1,045 *yds.* on Satur-day; how many yards did it turn out in the week?

Ans. 5,784 *yds.*

7. A merchant owes A. $2,160, B. $3,879, C. $813, D. $955, and E. $1,796; how much does he owe in all?

Ans. $9,603.

8. A farmer has 12 horses, 16 more oxen than horses, 42 more cows than horses and oxen together, and 22 more calves than oxen and cows together; how many in all?

9. A gentleman built a house; the carpenter work cost him $4,285, the masonry $3,950, the plumbing and grates $2,783, the painting $1,975, and miscellaneous work $3,992; what was the entire cost?

10. A merchant buys 56,250 *bu.* of corn, 30,211 *bu.* of oats, 18,312 *bu.* of barley, 2,197 *bu.* of wheat, and 713 *bu.* of rye; how many bushels did he buy altogether?

11. The distance from Albany to New York is 144 *miles,* from New York to Philadelphia 90 *miles,* from Philadelphia to Baltimore 98 *miles,* from Baltimore to Washington 38 *miles,* and from Washington to Norfolk 217 *miles;* how far is it from Albany to Norfolk by this route?

12. In a lumber yard there are 37,412 *ft.* of spruce, 15,102 *ft.* of pine, 9,187 *ft.* of oak, 171,812 *ft.* of hemlock, 7,413 *ft.* of ash, and 18,002 *ft.* of chestnut; how many feet are there of all kinds?

13. A work consists of 6 volumes; the first volume contains 611 pages, the second 539, the third 687, the fourth 599, the fifth 580, and the sixth 679; how many pages in the entire work?

14. A man bequeaths $15,750 to his daughter, $22,850 to each of two sons, and twice as much to his wife as to his daughter; what is the amount of his bequests?

15. The population of Maine is 627,413, of New Hampshire 301,471, of Vermont 300,187, of Massachusetts 1,240,499, of Connecticut 410,749, and of Rhode Island 192,815; what is the aggregate population of these States?

16. In 1876 the number of miles of railroad in the United States was as follows: in New England 5,694, in the Middle States 15,085, in the Western States 37,055, in the Southern States 16,676, and in the Pacific States 2,960; how many miles in all?

17. In 1876 the population of the several divisions of the United States was as follows: New England 3,806,850, Middle States 11,105,000, Western States 15,835,000, Southern States 12,410,000, Pacific States 1,280,395; what was the population of the entire country?

18. A merchant bought parcels of cloth containing respectively 3,912, 1,856, 2,011, 4,550, 937, 6,303, 1,856, 2,024, 4,228, 1,345, 6,138, 607, 960, 2,445, and 8,982 *yards;* how many yards did he buy in all?

19. The first of four numbers is 3,125, the second is greater than the first by 5,108, the third is equal to the sum of the first and second, and the fourth is equal to the sum of the third and first; what is the sum of the four numbers?

20. The ship Orient sailed from Marseilles to Buenos Ayres, distant 6,375 *miles*, thence to Valparaiso 2,764 *miles*, thence to San Francisco 6,346 *miles*, thence to the Sandwhich Islands 2,152 *miles*, thence to Melbourne 5,588 *miles*, thence to Yokohama 5,434 *miles*, thence to Calcutta 5,115 *miles*, thence to Bombay 2,257 *miles*, thence to Suez 2,006 *miles*, and thence back to Marseilles 1,314 *miles ;* what was the entire distance sailed ?

21. A merchant commenced business with the following capital : Cash $18,471.25, goods worth $21,419.52, bank stock $7,418.00, and other property worth $4,314.17 ; he gained $12,315.42 the first year, and $11,124.86 the second year ; how much was he worth at the end of the second year ?

22. An agent collected from different individuals: $27.18, $32.52, $41.70, $3.49, $8.17, $91.94, $127.86, $14.54, $87.78, $411.10, and $79.62 : how much did he collect in all ?

23. A man has real estate worth $20,114.50, bank stock worth $15,779.82, United States bonds worth $17,772.89, and other property worth $6,317.27 ; what is the value of his entire property ?

24. Find the sum of the following items of account : $21.27, $49.18, $412.25, $44.74, $86.92, $311.10, $8.14, $118.45, $32.41, and $52.52.

REVIEW QUESTIONS.

(**15.**) What is addition ? (**16.**) What is the sum of two or more numbers? What numbers can be added? (**17.**) When are numbers similar? Illustrate. (**18.**) Explain the use of the signs of addition and of equality. (**19.**) What are the principles of addition ? (**20.**) Give the rule for addition. The method of proving addition. (**21.**) What is a problem? The solution of a problem?

II. SUBTRACTION.

DEFINITIONS.

22. **Subtraction** is the operation of finding the difference between two numbers.

23. The **Difference** between two numbers is a number which, added to the *less*, will produce the *greater*. Thus, 6 is the difference between 10 and 4, because $4 + 6 = 10$.

The greater number is called the **Minuend**; the less number is called the **Subtrahend**; and their difference is called the **Remainder**.

The *minuend*, the *subtrahend*, and the *remainder* must be similar.

MENTAL EXERCISES.

1. James has 9 *marbles* and Samuel has 4 *marbles ;* how many more *marbles* has James than Samuel? If 4 marbles are taken from 9 marbles, how many will be left? 4 from 9 leaves how many?

2. John had 9 *chestnuts* and ate 6; how many had he left?   6 *chestnuts* from 9 *chestnuts* leaves how many *chestnuts?*   6 from 9 leaves how many?

3. Henry had 15 *cents,* but spent 9 *cents ;* how many *cents* had he left?   9 from 15 leaves how many?   What is the difference between 15 and 9?   16 and 9?   18 and 9?

4. William had 14 *apples,* of which he ate 3 and gave away 5; how many had he then?   What is the difference between 14 *apples* and the sum of 3 *apples* and 5 *apples?* 8 from 14 leaves how many?   8 from 17?   8 from 19? What is the difference between 14 and $3 + 5$?

EXPLANATION OF SIGNS.

24. The **sign of subtraction**, —, is called **minus** ; when placed between two numbers, it shows that the *second* is to be subtracted from the *first*. Thus, 5 — 3 shows that 3 is to be subtracted from 5.

A **parenthesis**, ( ), inclosing two or more numbers, shows that the inclosed expression is to be treated as a single number. Thus, 8 — (5 — 3) shows that the difference between 5 and 3 is to be subtracted from 8.

MENTAL EXERCISES.

1. What is the difference between 16 — 4 and 10 ?

2. What is the difference between 16 and 10 + 4 ?

3. 20 *sheep* + 2 *sheep* — (4 *sheep* + 7 *sheep*) = how many *sheep* ?

4. 19 *balls* — (12 *balls* — 6 *balls*) = how many *balls?*

5. 17 — 9 = how many ?

6. (17 + 10) — (9 + 10) = how many ?

7. 15 — 7 = how many ?

8. (15 + 10) — (7 + 10) = how many ?

9. $15 — $9 = how many dollars ?

10. $15 + $9 — ($3 + $4) = ?

11. 24 *lbs.* — (8 *lbs.* — 3 *lbs.*) = ?

12. (24 *lbs.* + 10 *lbs.*) — (5 *lbs.* + 10 *lbs.*) = ?

13. (20 + 16) — (10 + 10 — 9) = ?

14. (30 + 17) — (10 + 10 — 8) = ?

15. (40 + 15) — (30 + 7) = ?

PRINCIPLES OF SUBTRACTION.

25. The operation of subtraction depends on the following principles:

1°. *If all the parts of the subtrahend are taken from corresponding parts of the minuend, the sum of the partial remainders is equal to the required remainder.*

2°. *If the same number is added to both minuend and subtrahend, their difference is not changed.*

OPERATION OF SUBTRACTION.

26. Let it be required to find the difference between 565 and 393.

EXPLANATION.—We write the subtrahend under the minuend, so that units of the same order shall stand in the same column. Then, beginning at the right hand, we see that 3 *units* from 5 *units* leaves 2 *units;* we therefore write 2 in the

OPERATION.

Minuend,	565
Subtrahend,	393
Remainder,	172

line below. Because 9 *tens* cannot be taken from 6 *tens,* we increase the latter by 10 *tens,* making it 16 *tens;* now 9 *tens* from 16 *tens* leaves 7 *tens;* we therefore write 7 in the line below. To compensate for the 10 *tens,* or 1 *hundred* added to the minuend, we may *diminish* the 5 *hundreds* of the minuend by 1 *hundred,* or what is the same thing, we may increase the 3 *hundreds* of the *subtrahend* by 1 *hundred* (Principle 2°), which gives 4 hundreds ; taking 4 *hundreds* from 5 *hundreds,* we have 1 *hundred,* which we write in the line below. The resulting number, 172, is the sum of the partial remainders obtained by subtracting the parts of the subtrahend from corresponding parts of the minuend; it is, therefore, from Principle 1°, the required remainder.

In like manner we may find the difference between any two numbers ; hence, we have the following

R U L E .

I. Write the less number under the greater, so that units of the same order shall stand in the same column.

II. Beginning at the right, subtract each figure in the lower line from the one above it, and write the difference in the line below.

III. If any figure in the lower line exceeds the one above it, increase the latter by 10, perform the subtraction, and then add 1 to the next figure in the lower line.

The operation described in the last clause of the preceding rule is called *carrying.* This operation, and that of adding 10, when required, are performed mentally.

EXAMPLES.

	(1.)	(2.)	(3.)	(4.)
From	663	976 *lbs.*	704 *ft.*	1,806 *yds·*
Subtract	580	531 *lbs.*	483 *ft.*	720 *yds.*
Remainder,	83	445 *lbs.*	221 *ft.*	1,086 *yds.*

	(5.)	(6.)	(7.)	(8.)
From	4,236	80,502	$46,095	$555,555
Subtract	3,089	38,672	$28,736	$123,456
Remainder,	1,147	41,830	$17,359	$432,099

PROOF.—*Add the remainder to the subtrahend ; if the sum is equal to the minuend, the work is correct.*

ILLUSTRATIONS.

From	75,625	376,781 *lbs.*	$367,045	$84.16
Subtract	24,319	95,845 *lbs.*	$106,253	$29.18
Remainder,	51,306	280,936 *lbs.*	$260,792	$54.98
Proof,	75,625	376,781 *lbs.*	$367,045	$84.16

What is the difference between

13. 30,811 and 13,240 ?
14. 27,880 and 9,226 ?
15. 35,846 and 12,829 ?
16. 75,901 and 17,980 ?
17. 37,229 and 17,991 ?
18. 892,201 and 300,998 ?
19. 900,000 and 233,333 ?
20. 880,002 and 801,998 ?
21. 900,892 and 395 ?
22. 516,315 and 211,209 ?

23. 100,304 and 62,818 ?   25. 758,901 and 349,806 ?

24. 900,302 and 788,772 ?   26. 561,915,435 and 9,435 ?

27. The sum of two numbers is 7,817,412, and one of the numbers is 7,212,494; what is the other number ?

28. The greater of two numbers is 230,011 and the less is 210,299; what is their difference ?

29. The sum of two numbers is 485,752, and the less number is 82,992; what is the greater ?

30. What is the difference between 40,690,080 and 699,090 ?

Perform the following indicated subtractions:

31. 81,423 — 20,120.   39. $57,846,203 — $7,756.

32. 80,200 — 1,875.   40. $4,396,802 — $83,846.

33. 18,714 — 13,392.   41. 3,718,412 — 807,306.

34. 123,387 — 94,816.   42. 887,892 — 709,378.

35. $814,316 — $91,320.   43. 68,893 — 29,394.

36. $620,306 — $413,314.   44. 4,924,863 — 43,989.

37. $813,864 — $11,899.   45. 2,814,316 — 999,007.

38. 41,336 *yds.*— 7,814 *yds.*   46. 8,904,306 — 304,216.

47. From 2,816,214 *ft.* subtract 1,856,394 *ft.*

48. How much does 3,816,204 exceed 3,334,599 ?

49. What is the difference between 740,817 and 220,198 ?

50. From the sum of 862,141 and 32,843 subtract 884,109.

51. How much does the sum of 39,418 and 27,362 exceed the sum of 19,823 and 29,819 ?

52. Find the sum of 18,814 and 32,315, and subtract from it 17,794.

53. From the sum of $8,833, $141,209, and $11,362, subtract the sum of $2,843, and $10,906.

54. From the sum of 88,303 *feet*, and 67,119 *feet* subtract the sum of 74,395 *feet*, and 6,202 *feet*.

55. From 165,242 + 522,801 subtract 131,144 + 211,746.

56. From $21.56 + $42.87 + $11.72 subtract $48.99 + $2.65.

57. From $2,117.24 subtract $214.29 + $119.94 + $1.88.

58. From $38,140.20 subtract $16,884.49 + $22.27 + $46.71.

59. $4,547.18 + $1,620.29 — ($459.94 + $100.87 + $1,-257.00) = ?

60. $88,641 + $316.45 — ($19,384.22 — $6,211.88) = ?

PRACTICAL PROBLEMS.

1. A. borrowed of B. $9,780 and paid $2,176; how much remained due?　　　　　　　　　*Ans.* $7,604.

2. A. purchased a farm for $10,000 and paid thereon $4,790; how much remained due?　　　*Ans.* $5,210.

3. B. bought merchandise, which he sold for $11,275, and made thereby $2,114; what was the cost price?
　　　　　　　　　　　　　　　Ans. $9,161.

4. In 1860 the population of Maine was 627,413, and in 1870 it was 913,279; what was the gain in 10 years?
　　　　　　　　　　　　　　　Ans. 285,866.

5. The sum of two numbers is 9,427, and the greater is 5,825; what is the less number?

6. In 1790 the population of Connecticut was 238,141, and in 1840 it was 309,978; what was the gain in that period?

7. In 1840 the population of Arkansas was 97,574, which was a gain of 67,186 in 10 years; what was the population of that State in 1830?

2

8. How much does 57,182 exceed 18,394 ?

9. A merchant commenced business with a capital of $21,308, and retired with $74,114; how much did he make?

10. A., B., and C., commence business; A. puts in $35,000, B. $41,700, and C. $36,150; at the end of a year they have together $149,711 : how much did they gain?

11. A merchant bought 500 yards of linen for $276, 3,400 yards of muslin for $325, and 75 yards of broad-cloth for $318, and sold the whole for $1,316; how much did he gain?

12. A. has a yearly income of $12,000; of this he spends for rent $2,750, for taxes, repairs, and insurance, $814, for clothing $1,342, for household expenses $6,211, and the remainder he distributes in charity: how much does he distribute?

14. B. has $12,311, and after paying his debts and giving away $2,108, he has remaining $8,199 ; what is the amount of his debts?

14. A merchant bought cloth for $1,592, silk for $1,274, laces for $818, and sold the cloth for $2,102, the silk for $1,190, and the laces for $969 ; how much did he gain?

15. A landholder owned 1,875 acres in Illinois, 2,396 acres in Indiana, and 13,742 acres in Michigan; of this he sold 813 acres in Illinois, 372 acres in Indiana, and 7,411 acres in Michigan : how many acres has he remaining?

16. A., B., and C., are in trade; A. gains $7,055, B. gains $813 less than A., and C. gains as much as A. and B. together, lacking $994 : what do they all gain?

17. A. bought a farm for $8,192, expended $2,815 for improvements, paid $387 for taxes, and then sold it so as to lose $2,282 ; for what did he sell it ?

18. A man worth $18,000 left $4,287 to his elder son, $3,754 to his younger son, $3,219 to his daughter, and the remainder to his wife ; what was the wife's portion ?

19. A man was 21 years old in 1843 ; in what year will he be 75 years old ?

20. A merchant bought 4 cargoes of grain ; the *first* contained 6,705*bu.*, the *second* contained 842*bu.* less than the first, the *third* contained 911*bu.* more than the second, and the *fourth* contained 3,092*bu.* less than the second and third together : how many bushels were there in the four cargoes ?

21. A man bought three estates ; for the *first* he gave $5,260, for the *second* he gave $3,585, and for the *third* he gave as much as for the first two together ; he afterward sold them all for $15,280 : did he gain or lose, and how much ?

22. A. travels due east at the rate of 19 miles an hour ; B. starts from the same place 1 hour later and travels in the same direction at the rate of 13 miles an hour ; how far apart are they 3 hours after A. starts ?

23. A. travels due north at the rate of 17 miles an hour ; B. starts from the same place an hour earlier, and travels due south at the rate of 11 miles an hour ; how far apart are they 4 hours after A. starts ?

24. A merchant commenced business, having in cash $4.152,17, in goods $11,443.12, and in other property $5,794.22 ; at the end of a year he had in cash $2,158.23,

in goods $17,411.98, and in other property $6,239.14:
how much did he gain in the year?

25. A gentleman purchased a house for $12,873.75, a
carriage for $720.50, a span of horses for $591.45, and a
saddle horse for $212;25; he paid for them at one time
$4,374.16, at another time $3,495.17, and at a third time
$2,675.14: how much remained unpaid?

26. The areas of the New England States are as follows:
Maine has 30,408 square miles, New Hampshire has 9,386,
Vermont 9,420, Massachusetts 7,845, Connecticut 4,693,
and Rhode Island 1,395; how many fewer square miles
has Maine than all the rest together?

27. A drover bought 24 oxen for $1,214.26, 42 cows for
$2,111.79, and 40 calves for $397.11; he sold the oxen for
$1,519.45, the cows for $2,237.18, and the calves for
$318.27; what did he gain by the transaction?

28. America was discovered in 1492, which was 128 years
before the settlement of New England; in what year was
New England settled?

29. A man having a sum of money, earned $8,211, and
afterward lost $2,114, when he found that he had $11,415;
how much had he at first?

30. In a division there were 11,376 men, of whom 696
were killed in battle; how many remained?

REVIEW QUESTIONS.

(22.) What is subtraction? (23.) What is the difference between
two numbers? Illustrate. What is the minuend? The subtrahend?
The remainder? (24.) What is the name and use of the sign of
subtraction? What is the use of the parenthesis? (25.) What
are the principles of subtraction? (26.) Give the rule for subtraction. How is subtraction proved?

III. MULTIPLICATION.

DEFINITIONS.

27. Multiplication is the operation of taking one number as many times as there are units in another.

The first number, or the number to be repeated, is called the **Multiplicand**; the second number is called the **Multiplier**; and the result is called the **Product**. Thus, 4 *multiplied* by 3 is equal to $4+4+4$, or to 12. Here 4 is the *multiplicand,* 3 is the *multiplier,* and 12 is their *product.*

Both *multiplicand* and *multiplier* are called **Factors** of the *product.* Thus, 4 and 3 are *factors* of 12.

MENTAL EXERCISES.

1. What is the cost of 3 oranges at 6 *cents* apiece?

EXPLANATION.—Because 1 orange costs 6 *cts.*, 3 oranges will cost 6 *cts.* + 6 *cts.* + 6 *cts.*, or 18 *cts.* From this we see that multiplication is a short method of performing the operation of addition when the numbers to be added are equal to each other.

2. A farmer sells 6 calves at \$9 each; how much does he receive? $\$9+\$9+\$9+\$9+\$9+\$9 = ?$ How much is 6 times \$9? 6 times 9?

3. If a man earns \$5 a day, how much will he earn in 9 days? What is 9 times \$5? What is the product of 9 and 5?

4. How many are 4 times 5? How many are 5 times 4?

EXPLANATION.—The product does not depend on the order of the factors, as may be seen in the diagram. If we take the stars by columns we have 4 times 5 *stars;* if we take them by horizontal rows, we have 5 times 4 *stars;* in either case, we have 20 *stars.*

```
* * * *
* * * *
* * * *
* * * *
* * * *
```

5. How many are 9 times 8? 8 times 9? 4 times 10? 7 times 12? 12 times 12?

28. The **Sign of Multiplication**, ×, when placed between two numbers, indicates that their product is to be taken. Thus, the expression 5 × 7 shows that 5 is to be *multiplied by* 7, or that 7 is to be *multiplied by* 5.

CONDENSED MULTIPLICATION TABLE.

1	2	3	4	5	6	7	8	9	10	11	12
2	4	6	8	10	12	14	16	18	20	22	24
3	6	9	12	15	18	21	24	27	30	33	36
4	8	12	16	20	24	28	32	36	40	44	48
5	10	15	20	25	30	35	40	45	50	55	60
6	12	18	24	30	36	42	48	54	60	66	72
7	14	21	28	35	42	49	56	63	70	77	84
8	16	24	32	40	48	56	64	72	80	88	96
9	18	27	36	45	54	63	72	81	90	99	108
10	20	30	40	50	60	70	80	90	100	110	120
11	22	33	44	55	66	77	88	99	110	121	132
12	24	36	48	60	72	84	96	108	120	132	144
13	26	39	52	65	78	91	104	117	130	143	156
14	28	42	56	70	84	98	112	126	140	154	168
15	30	45	60	75	90	105	120	135	150	165	180
16	32	48	64	80	96	112	128	144	160	176	192
17	34	51	68	85	102	119	136	153	170	187	204
18	36	54	72	90	108	126	144	162	180	198	216
19	38	57	76	95	114	133	152	171	190	209	228
20	40	60	80	100	120	140	160	180	200	220	240

USE OF THE TABLE.—Find the multiplicand in the upper line and the multiplier in the first column; their product will then be found in the same column with the multiplicand and in the same line with the multiplier, By reversing the process, any number given in the table may be separated into factors. Thus, $187 = 17 \times 11$.

MENTAL EXERCISES.

1. Add by 2's from 2 to 40. By 3's from 3 to 60. By 4's from 4 to 80. By 5's from 5 to 100.

2. Subtract by 2's from 40 to 2. By 3's from 60 to 3. By 4's from 80 to 4. By 5's from 100 to 5.

NOTE.—Let these exercises be continued to the limit of the table.

3. What is the cost of 12 *lbs.* of sugar at 15 *cts.* a pound ? 12 times 15 *cts.* are how many cents ? What is the product of 15 *cts.* by 12 ? Of 15 by 12 ?

4. A man earns $18 a week, and it costs him $11 a week to live ; how much can he save in 13 weeks ? in 15 weeks ? in 9 weeks ? ($18—$11) $\times 13 =$? $(18-11) \times (14-6) =$?

NOTE.—The multiplier must always be *abstract ;* the multiplicand may be either *abstract,* or *denominate ;* the product is always *similar* to the multiplicand. In practice, we multiply as though both factors were abstract, and then determine the unit of the product from the nature of the question.

5. $(26-9) \times 8 =$?

6. $(24-13) \times (15-7) =$?

7. $(20-7) \times (19-8) =$?

8. $(3+9) \times (11+6) =$?

9. $8 \times 10 - 14 =$?

10. $4 \times 9 - 13 \times 2 =$?

11. $(4+9) \div (18-7) =$?

12. $(7+10) \times (12-3) =$?

MULTIPLICATION BY ONE FIGURE.

29. Let it be required to multiply 946 by 8.

EXPLANATION.—Multiplying 6 *units* by 8, we have 48 *units ;* that is, 4 *tens* and 8 *units ;* we set down 8 in the units' place, and carry the 4 *tens* to the next column. Multiplying 4 *tens* by 8 we have 32 *tens,* which increased by the 4 *tens* brought forward give 36 *tens,* or 3 *hundreds* and 6 *tens ;* we set down 6 in the tens' place, and carry

OPERATION.

Multiplicand,	946
Multiplier,	8
Product,	7,568

the 3 *hundreds* to the next column. Multiplying 9 *hundreds* by 8 and adding the 3 *hundreds* brought forward, we have 75 *hundreds,* which we set down. The resulting number 7,568 is the required product.

In like manner we may proceed in all similar cases; hence, the following

RULE.

Begin at the right hand and multiply each figure of the multiplicand by the multiplier, setting down and carrying as in addition.

NOTE.—We may multiply by any number from 10 to 20 by the same rule.

EXAMPLES.

Perform the following multiplications:

	(1.)	(2.)	(3.)	(4.)
Multiplicand,	357	8645	2079	84123
Multiplier,	5	8	9	6
Product,	1785	69160	18711	504738

(5.)	(6.)	(7.)	(8.)
8842 *yds.*	3749 *in.*	13146 *lbs.*	\$81386
4	7	9	8
35368 *yds.*	26243 *in.*	118314 *lbs.*	\$651088

(9.)	(10.)	(11.)	(12.)
56432	13596 *ft.*	14382 *lbs.*	\$87645
12	15	17	19

13. $43875 \times 9 = ?$

14. $14876 ft. \times 11 = ?$

15. $\$87653 \times 14 = ?$

16. $79792 lbs. \times 12 = ?$

17. $16749 \times (24-7) = ?$

18. $86639 \times 12 = ?$

19. $\$39864 \times 18 = ?$

20. $\$222794 \times 19 = ?$

21. $637489 \times 9 = ?$

22. $333333 \times 16 = ?$

EFFECT OF ANNEXING CIPHERS.

30. Every cipher that we annex to a number moves each of its digits one place to the left, that is, it converts *units* into *tens*, *tens* into *hundreds*, and so on; but this is . the same as multiplying the number by 10; hence,

To multiply a number by 10, we annex one cipher; to multiply it by 100, we annex two ciphers; to multiply it by 1000, we annex three ciphers; and so on.

Thus, $75 \times 10 = 750$; $75 \times 100 = 7,500$; $75 \times 1000 = 75,000$; $75 \times 10,000 = 750,000$; and so on.

To multiply by *any number of tens*, we first multiply by the *given number* and then annex *one* cipher to the product; to multiply by *any number of hundreds*, we multiply by the *given number* and annex *two* ciphers; to multiply by *any number of thousands*, we multiply by the *given number* and annex *three* ciphers; and so on. Thus, 8×4 *tens* $= 32$ *tens* $= 320$; 6×4 *hundreds* $= 24$ *hundreds* $= 2,400$; $7 \times 3,000 = 21,000$; and so on.

PRINCIPLES OF MULTIPLICATION.

31. The operation of multiplication depends on principles already explained and also on the following:

If all the parts of the multiplicand are multiplied by each part of the multiplier, the sum of the partial products is equal to the required product.

MULTIPLICATION BY ANY NUMBER.

32. Let it be required to find the product of 458 and 346.

EXPLANATION.—Having written the numbers so that units of the same order stand in the same column, we begin at the right and multiply all the parts of the multiplicand by 6, as explained in Art. 29; this gives 2748 for the first partial product.

We next multiply all the parts of the multiplicand by 4 *tens*, or 40.

OPERATION.

Multiplicand,	458
Multiplier,	346
	2748
Partial products,	1832
	1374
Product,	158468

Multiplying 8 *units* by 40 (Art. **30**), we have 320, that is, 3 *hundreds* and 2 *tens;* we omit the cipher, write 2 in the column of tens, and carry 3 to the column of hundreds, and so on ; this gives the second partial product.

We next multiply all the parts of the multiplicand by 3 *hundreds.* Multiplying 8 *units* by 300 we have 2400, or 2 *thousands* and 4 *hundreds;* we omit the two ciphers, write 4 in the column of *hundreds,* and carry 2 to the column of *thousands,* and so on ; this gives the third partial product.

The sum of the products thus obtained is 158,468; but this is *the sum of the partial products found by multiplying all the parts of the multiplicand by each part of the multiplier;* it is therefore the required product (Art. **31**).

In like manner we may find the product of any two numbers ; hence, the following

R U L E .

I. Write the multiplier under the multiplicand, so that units of the same order shall stand in the same column.

II. Beginning at the right, multiply all the parts of the multiplicand by each figure of the multiplier, writing the first figure of each partial product under the corresponding multiplier.

III. Find the sum of the partial products.

EXAMPLES.

(1.)	(2.)	(3.)	(4.)	(5.)
843	1817	7287	325	9372
27	69	75	503	98
5901	16353	36435	975	74976
1686	10902	51009	1625	84348
22761	125373	546525	163475	918456

PROOF.—*Multiply the multiplier by the multiplicand ; if the product is the same as before, the work is correct.*

(6.)	*Proof.*	(7.)	*Proof.*
345	572	835	794
572	345	794	835
690	2860	3340	3970
2415	2288	7515	2382
1725	1716	5845	6352
197340	197340	662990	662990

Multiply

8. 875 by 349.

9. $1,843 by 216.

10. 1,781 by 74.

11. 939 by 77.

12. 1,754 by 306.

13. 7,506 by 45.

14. 2,016 by 1,008.

15. 8,435 by 371.

16. 4,572 by 614.

17. 32,183 by 179.

18. 46,137 by 841.

19. 50,246 by 322.

20. 61,532 by 742.

21. 184,387 by 994.

22. 97,418 by 887.

23. 107,309 by 1,206.

24. 320,009 by 344.

25. 99,897 by 284.

26. 44,479 by 3,227.

27. 56,263 by 7,777.

When the multiplier terminates in ciphers, multiply the significant part, and annex the ciphers as explained in Article **30.**

(28.)	(29.)	(30.)	(31.)	(32.)
31	875	42	87	3194
290	4300	31000	1500	127000
8990	3762500	1302000	130500	405638000

33. 87 by 78,000.

34. 314 by 87,000.

35. 414 by 82,000.

36. 4,968 by 3100.

37. 19,872 by 26000.

38. 346,843 by 4500.

If both multiplicand and multiplier terminate in ciphers, multiply the significant parts, and annex as many ciphers as there are in both factors.

39. 8,840 by 7,250.
40. 2,040 by 8,060.
41. 10,800 by 870.
42. 37,300 by 8,170.
43. 88,320 by 36,000.
44. 45,100 by 8,190.

45. 32,400 by 32,400.
46. 18,750 by 16,000.
47. 37,590 by 92,000.
48. 33,330 by 27,100.
49. 877,000 by 209,000.
50. 337,800 by 99,000.

51. $3,876 \times 879 - 2,799 \times 74 = ?$
52. $12,483 \times 4,520 - 38,795 \times 89 = ?$
53. $1,379 \times 794 + 145,902 \times 86 = ?$
54. $(14,749 - 3,892) \times (12,700 - 8,309) = ?$
55. $(265,484 - 142,184) \times (8,794 - 3,684) = ?$
56. $(18,943 + 37,711) \times (27,385 - 7,965) = ?$
57. $(7,890 + 8,901) \times (3,700 + 6,400) = ?$
58. $(276 + 3,276) \times 875 - 4,962 \times 79 = ?$
59. $(1,846 - 199) \times 79 - (1,329 - 211) \times 9 = ?$
60. $(5,946 + 9,544) \times 284 + (8,305 + 95) \times 890 = ?$

ADDITIONAL DEFINITIONS.

33. A **Continued Product** is a product of more than two factors. Thus, $3 \times 4 \times 5$ is a continued product; it indicates that the product of 3 and 4 is to be multiplied by 5.

A continued product may contain any number of factors. Its value is independent of the order of the factors. Thus, $3 \times 4 \times 5 = 3 \times 5 \times 4 = 4 \times 3 \times 5$.

34. A **Composite Number** is a number composed of two or more integral factors. Thus, 21 is a composite number, because it is equal to 3×7; the number 30 is composite, because it is equal to $2 \times 3 \times 5$.

To multiply a number by a composite number whose factors are known we have the following

R U L E .

Multiply the multiplicand by one factor of the multiplier, then multiply the result by another factor, and so on, till all the factors have been used.

E X A M P L E S .

1. Multiply 324 by 36, that is, by 9×4, or by 12×3.

FIRST OPERATION.		SECOND OPERATION.	
324		324	
9		12	
2916	Partial product.	3888	Partial product.
4		3	
11664	Total product.	11664	Total product.

NOTE.—In the following examples the factors of the multiplier may be found by means of the multiplication table. If the factors are unequal, we generally begin with the greater one.

2. $873 \times 144 = ?$ 7. $736 \times 48 = ?$

3. $887 ft. \times 84 = ?$ 8. $4,315 \times 176 = ?$

4. $\$3,845 \times 63 = ?$ 9. $\$8,712 \times 209 = ?$

5. $38,257 yds. \times 96 = ?$ 10. $48 \times 11 \times 15 \times 16 = ?$

6. $7,836 lbs. \times 132 = ?$ 11. $234 \times 14 \times 12 \times 7 = ?$

PRACTICAL PROBLEMS.

1. What will 455 *lbs.* of sugar cost at 14 cents per pound ?   *Ans.* 6370 *cts.* = \$63.70.

2. What will 692 pounds of beef cost at 26 cents per pound ?   *Ans.* 17,992 *cts.* = \$179,92.

3. If one barrel of pork costs \$15, what will 1,728 barrels cost?   *Ans.* \$25,920.

4. If a train travels 35 miles an hour, how far will it travel in 425 hours?   *Ans.* 14,875 miles.

5. An army contains 106 regiments, and each regiment contains 1,128 men; how many men in the army?

6. If it requires 720 barrels of provisions to feed an army for one day, how many barrels will it require for 365 days?　　　　　　　　　　*Ans.* 262,800 barrels.

7. There are 320 rods in a mile; how many rods are there in 50 miles?

8. If a railway costs $42,500 per mile, and is 385 miles long, how much does it cost?

9. A field containing 56 acres produces 29 *bu.* of rye to the acre; what is the total yield?

10. Sound travels at the rate of 1,142 feet per second; how far will it travel 3,600 seconds, or one hour?

11. The distance from New York to Bridgeport is 56 miles, and there are 320 rods in a mile; how many rods from New York to Bridgeport?

12. In an orchard there are 214 rows of trees, and each row contains 241 trees; how many trees are there in the orchard?

13. What is the continued product of 92, 37, and 45?

14. A freight train consists of 21 cars; each car contains 85 barrels of flour, and each barrel of flour weighs 196 pounds: how many pounds in the entire cargo?

15. The distance from Bridgeport to New Haven is 18 miles; each mile contains 1,760 yards, and each yard 3 feet: how many feet from Bridgeport to New Haven?

16. In an orchard there are 14 rows of peach trees; each row contains 27 trees, and each tree bears 108 peaches: how many peaches in the orchard?

17. A man earns $18.75 a week, and all his expenses are $11.25 a week; how much can he save in 27 weeks?

18. A woman bought 18 *yds.* of ribbon at 27 *cts.* a yard, and 42 *yds.* of muslin at 16 *cts.* a yard; what did she pay for both?

19. A man has a barn worth $475, a house worth 5 times as much as the barn, and land worth 3 times as much as the house and barn together; what are they all worth?

20. A man traveled 295 miles in 6 days; for 5 days he traveled at the rate of 53 miles a day: how far did he travel the sixth day?

21. The diameter of Mercury is 2,967 miles; the diameter of Saturn is 24 times that of Mercury; and the diameter of the sun is 12 times that of Saturn: what is the sun's diameter?

22. A courier had to travel a certain distance in 13 hours; for 5 hours he traveled at the rate of 12 miles an hour, but finding that he was behind time, he increased his speed and finished the journey at the rate of 14 miles an hour: what was the distance traveled?

23. The distance from Chicago to Albany is 835 miles; a passenger train starts from Chicago and runs toward Albany at the rate of 38 miles an hour, and at the same time a freight train starts from Albany and runs toward Chicago at the rate of 13 miles an hour: how far apart are they at the end of 12 hours?

24. A steamboat runs from St. Louis to Cairo in 11 hours at the rate of 17 miles an hour; from Cairo to Memphis in 15 *hrs.*, at the rate of 16 *mi.* an hour; from Memphis to Vicksburgh in 23 *hrs.*, at the rate of 18 *mi.* an hour; and

from Vicksburgh to New Orleans in 21 *hrs.*, at the rate of 19 *mi.* an hour: what was her running distance from St. Louis to New Orleans ?

25. A fox starts from a certain place and runs at the rate of 616 *yds.* a minute; at the end of three minutes a dog starts from the same place and follows the fox at the rate of 792 *yds.* a minute: how far apart are they at the end of 9 minutes?

REVIEW QUESTIONS.

(**27.**) What is multiplication? Multiplicand? Multiplier? Product? Define factors. (**28.**) Explain the use of the sign of multiplication. (**29.**) Give the rule for multiplying by a single figure. (**30.**) How do you multiply a number by 10? By 100? By 1000? By any number of tens? (**31.**) What is the fundamental principle of multiplication? (**32.**) Give the rule for multiplication by any number. If the multiplicand is denominate, what will be the nature of the product? What is the method of proving multiplication? What is the rule for multiplying when both factors terminate in ciphers? (**33.**) What is a continued product? How many factors may such a product contain? (**34.**) What is a composite number? How do you multiply by a composite number whose factors are known?

IV. DIVISION.

DEFINITIONS.

35. Division is the operation of finding how many times one number is contained in another, or of finding one of the equal parts of a number.

The number to be divided is the **Dividend** ; the number by which it is divided is the **Divisor** ; the result of the division is the **Quotient** ; and the part of the dividend that remains after the operation is the **Remainder.**

When the remainder is 0, the division is said to be **exact**; in this case both *the divisor* and *the quotient* are *factors* of the dividend.

MENTAL EXERCISES.

1. If 24 *apples* are divided equally among 6 boys, how many *apples* will each boy receive? What is one of the 6 equal parts of 24 *apples?* How many times is 6 contained in 24?

NOTE.—To use the *multiplication table* or a *division table*, find the divisor in the left hand column, and on the same line find the dividend; the quotient will be the number at the head of the corresponding column. Let the pupil familiarize himself with this method of using the table.

2. What is the quotient of 36 divided by 6? of 144 by 12? of 126 by 14? of 72 by 9? of 153 by 17? of 90 by 18? of 95 by 19? of 56 by 14?

3. How many 7's can be taken from 23, and what will remain? What is the quotient of 23 divided by 7, and what is the remainder? Subtract by 7's from 23 as far as possible, and find the remainder.

NOTE.—Division may be regarded as a short method of continued subtraction. The *number of times* that the divisor can be taken from the dividend is equal to the *quotient.*

4. If 47 *peaches* are divided into 9 equal parcels, how many will there be in each parcel, and how many over? In this case, what is the *dividend?* The *divisor?* The *quotient?* The *remainder?*

5. If you divide 126 by 11, what is the quotient, and what is the remainder? 149 by 12? 74 by 6? 190 by 15?

6. If you divide $154 among 14 children, how much will each child receive? What is the quotient of 154 divided by 14? How many times is 14 contained in 154.

Note.—Division is performed as though both numbers were abstract and the unit is determined from the nature of the question. In the last example $1.54 is equal to 154 *cts.*; we divide 154 by 14, which gives 11 ; hence, the answer is 11 *cts.*

If the dividend and divisor are *similar* the quotient is abstract ; if the dividend is *denominate*, the quotient is of the same denomination as the dividend.

7. How many yards of cloth, at $7 a yard, can be bought for $84 ? If 12 *yds.* of cloth cost $84, what is the cost of 1 *yd.?* How many times is 7 contained in 84 ? What is the quotient of 84 by 7 ? of 84 by 12 ?

METHODS OF INDICATING DIVISION.

36.—1°. The **sign of division,** $\div$, when placed between two numbers, indicates that the *first* is to be divided by the *second.* Thus, the expression $12 \div 3$ indicates that 12 is to be divided by 3.

2°. The same operation may be indicated by writing the dividend over the divisor, with a line between them. Thus, the expression $\frac{12}{3}$, which is read 12 *divided* by 3 is equivalent to the expression $12 \div 3$.

3°. Division may also be indicated by writing the divisor on the left of the dividend, with a curved line between them. Thus, the expression $3) 12$ is equivalent to the expression $\frac{12}{3}$, or to $12 \div 3$.

MENTAL EXERCISES.

1. What is the quotient of 144 by 12 ? of $96 \div 12$? of $96 \div 12 ? of $84 \div 12$? How many times is $12 contained in $84 ?

2. $\frac{72}{9} \times \frac{54}{8} = $?

3. $190 \div 19 - 54 \div 6 = $?

4. $192 \div (23 - 7) = $?

5. $(200 - 8) \div 16 = $?

6. $(108 + 45) \div (10 + 7) = ?$

7. $(4 + 9 + 26) \div 13 = ?$ 9. $(12 \times 17) \div 6 = ?$

8. $156 \div (19 - 6) = ?$ 10. $(12 \times 16) \div (13 - 5) = ?$

11. $(99 \div 11) \times (154 \div 14) = ?$

12. The product of two numbers is 64 and one of the numbers is 16; what is the other? $64 \div 16 =$ how many?

13. Write the following by means of signs: the sum of $12 and $10, divided by $11, is equal to the quotient of the difference between $18 and $4, divided by $7.

NOTE.—If a thing is divided into 2 equal parts, each part is called **one half**; if divided into 3 equal parts, each is called **one third**; if into 4 each is **one fourth**; if into 5 each is **one fifth**; and so on. Thus, $\frac{1}{2}$ is *one half;* $\frac{3}{2}$ is *one half* of 3, or 3 *halves* of 1; $\frac{1}{3}$ is *one third;* $\frac{2}{3}$ is *one third* of 2, or *two thirds* of 1; $\frac{1}{4}$ is *one fourth;* $\frac{2}{4}$ is *one fourth* of 2, or 2 *fourths* of 1; $\frac{3}{4}$ is *one fourth* of 3, or 3 *fourths* of 1; and so on. Expressions of the form $\frac{1}{2}, \frac{1}{4}, \frac{2}{3}, \frac{3}{4}$, etc., are called **fractions.** Fractions are treated more fully hereafter.

14. What is $\frac{1}{2}$ of 4? $\frac{1}{2}$ of 8? $\frac{1}{2}$ of 16? $\frac{1}{3}$ of 3? $\frac{1}{3}$ of 12? $\frac{1}{3}$ of 15? $\frac{1}{3}$ of 9? $\frac{1}{4}$ of 4? $\frac{1}{4}$ of 8? $\frac{1}{4}$ of 64?

15. Read the expressions $\frac{3}{2}, \frac{4}{3}, \frac{7}{8}, \frac{9}{5}, \frac{7}{11}, 1\frac{4}{5}, 2\frac{2}{7}, 1\frac{8}{9}.$

OBJECT OF DIVISION.

37. **Division is the reverse of multiplication.** In multiplication, we have *two factors* given, to find their product; in division, we have the *product and one factor* given, to find the other factor.

PRINCIPLES OF DIVISION.

38. The operation of division depends on the principles obtained by reversing those of Article **30**, and also on the following, obtained by reversing that of Article **31**:

If we divide all the parts of the dividend by the divisor, the sum of the partial quotients is equal to the required quotient.

39. There are two cases: 1°. **Short Division,** in which the divisor contains but one figure; and, 2°. **Long Division,** in which the divisor contains more than one figure.

In the *first* case most of the operation is performed mentally ; in the second case the different steps of the operation are written out. The principles employed are the same in both cases.

I° CASE. SHORT DIVISION.

40. Let it be required to divide 26,812 by 4 :

EXPLANATION.—Having written the numbers as shown in the margin, we begin at the left and divide the different parts of the dividend by the divisor.

OPERATION.

Dividend.

Divisor, 4) 26812

Quotient,   6703

Since 2 is not divisible by 4, we divide 26 by 4 ; this gives 6 for a quotient with 2 for a remainder ; hence, there are 6 *thousands* in the quotient; we write 6 in the column of thousands and to the remainder we annex the following figure of the dividend giving 28 *hundreds*. The quotient of 28 by 4 is 7 ; hence, there are 7 *hundreds* in the quotient ; we therefore write 7 in the column of hundreds. Since there is no remainder and since 1 is smaller than 4 there are no *tens* in the quotient ; we therefore write 0 in the place of tens and annex the following figure to 1 giving 12 *units*. The quotient of 12 by 4 is 3, which we write in the column of units.

In like manner we may treat all similar cases ; hence, the following

RULE.

I. *Write the divisor on the left of the dividend, and draw a line between them.*

II. *Divide the first figure of the dividend by the divisor and set the quotient underneath, or, if the first figure is less than the divisor, divide the first two figures and set the quotient under the second.*

III. To the remainder annex the following figure of the dividend, divide the result by the divisor and set the quotient underneath, or, if the result is less than the divisor, put a cipher in the quotient, annex another figure, and proceed as before.

IV. Continue the operation till all the figures of the quotient have been found.

EXAMPLES.

Perform the following divisions:

	(1.)	(2.)	(3.)	(4.)	(5.)
	5)785	6)804	8)1624	7)392	9)1926
Ans.	157	134	203	56	214

	(6.)	(7.)	(8.)	(9.)	(10.)
	4)1544	3)825	8)4896	9)792	7)2415
Ans.	386	275	612	88	345

If there is a remainder after the last partial division, we write it over the divisor and annex the result to the quotient. Thus, $27 \div 4 = 6\frac{3}{4}$ indicates that the quotient of 27 by 4 is 6 with a remainder 3. The expression $6\frac{3}{4}$ may be read 6 and 3 *divided by* 4, or 6 and 3 *fourths*.

	(11.)	(12.)	(13.)	(14.)
	5)176	4)8140	7)8146	9)4023
Ans.	$35\frac{1}{5}$	2035	$1163\frac{5}{7}$	447

PROOF.—*Multiply the quotient by the divisor and to the product add the remainder; if the result is equal to the dividend, the work is correct.*

Thus, in example (11.), $35 \times 5 + 1 = 176$; hence, the work is correct.

Divide

15. 12,360 by 4.
16. 3,730 by 5.
17. 20,202 by 6.
18. 37,904 by 4.
19. 90,872 by 8;
20. 640,339 by 7.

Divide

21. $639,145 by 7.
22. $454,396 by $8.
23. 321,314 *lbs.* by 7.
24. 1,287,643 *ft.* by 9 *ft.*
25. $21,416,314 by 9.
26. 82,324,717 *yds.* by 8.

We may divide by any number from 10 to 20 in the same manner as when the number is expressed by one figure.

(27.)	(28.)	(29.)	(30.)
10)8760	11)10978 *ft.*	12)8134	11)203236 *mi.*
876	998 *ft.*	677$\frac{10}{12}$	18476 *mi.*

Divide

31. $888,888 by $12.
32. 77,077 *lbs.* by 11 *lbs.*
33. 6,809 *yds.* by 11 *yds.*
34. $26,736 by $12.
35. 154,824 by 12.
36. 14,784 *mi.* by 11 *mi.*

Divide

37. 137,418 by 12.
38. 124,212 by 12.
39. 89,580 by 15.
40. $24,311 by 16.
41. $33,244 by $15.
42. 403,210 by 12.

43. ($18,211 — $13,179) ÷ (19 — 14) = ?
44. ($8,444 — $4,514) ÷ ($22 — $15) = ?
45. ($53.92 — $7.27) ÷ 15 = ?
46. ($211 — $175) ÷ 16 = ?
47. (2,500 *lbs.* + 2,065 *lbs.*) ÷ 11 = ?
48. ($100 + $28.94) ÷ 14 = ?
49. ($100 — $21.52) ÷ 9 = ?
50. (5,000 *yds.* — 320 *yds.*) ÷ 10 = ?
51. ($76 — $25.84) ÷ 12 = ?
52. (893 *ft.* — 491 *ft.*) ÷ 6 = ?

2° CASE. LONG DIVISION.

41. Let it be required to divide 60,198 by 237.

EXPLANATION.—Having written the divisor on the left, we divide as explained below.

OPERATION.
Divisor. Dividend. Quotient.

$$237) 60198 (254$$
$$\underline{474}$$
$$1279$$
$$\underline{1185}$$
$$948$$
$$\underline{948}$$

Remainder, 0

Since 60 is not divisible by 237, there are no *thousands* in the quotient; annexing 1, we have 601 *hundreds*. The quotient of 601 by 237 is greater than 2, but less than 3 ; hence, there are 2 *hundreds* in the quotient ; multiplying 237 by 2 *hundreds*, we have 474 *hundreds*, and this taken from 601 *hundreds*, leaves 127 *hundreds*, to which we bring down and annex the 9 *tens*, giving 1,279 *tens*. The quotient of 1,279 by 237 is greater than 5 and less than 6 ; hence, there are 5 *tens* in the quotient ; multiplying 237 by 5 *tens*, we have 1,185 *tens*, and this taken from 1,279 *tens*, leaves 94 *tens*, to which we bring down and annex the 8 *units*, giving 948. The quotient of 948 by 237 is 4 ; hence, there are 4 units in the quotient ; multiplying 237 by 4, we have 948, which taken from 948, leaves 0.

` In like manner we may treat all similar cases ; hence, the following

RULE.

I. Find how many times the divisor is contained in the fewest possible figures on the left of the dividend for the first figure of the quotient ; multiply the divisor by this figure and subtract the product from the figures used.

II. Annex to the remainder thus found the next figure of the dividend, and find how many times the divisor is contained in the result for the second figure of the quotient ; multiply the divisor by this figure and subtract as before.

III. Continue this operation till all the figures of the quotient have been found.

In applying the preceding rule, it is found convenient to write the *divisor* on the *left* and the *quotient* on the *right* of the dividend. Should there be a remainder after the operation is completed, it is to be treated as explained in short division.

The method of proof, which is the same as for short division, is illustrated in the example below.

EXAMPLES.

Perform the following divisions:

1. Divide 35,114 by 143.

OPERATION.

Dividend.		PROOF.	
Divisor, 143)35114(245 Quotient.		245	Quotient.
286		143	Divisor.
651		735 ⎞	
572		980 ⎬ Partial products.	
794		245 ⎠	
715		79	Remainder.
79 Remainder.		35114	Dividend.

The practical method of adding the remainder to the product of the quotient and the divisor is indicated in the above example.

(2.)	(3.)	(4.)
36)7452(207	124)373116(3009	37)11248(304
72	372	111
252	1116	148
252	1116	148

In example (2), the partial dividend, 25, is smaller than the divisor; we therefore write 0 for the second figure of the quotient and bring down 2, the next figure of the dividend, and proceed as before.

In example (3), both 11 and 111 are smaller than the divisor; we therefore write two ciphers in the quotient, bring down 6, and proceed as before.

Divide as indicated :

5. $7,812 \div 36$.

6. $16,758 \div 49$.

7. $14,464 \div 64$.

8. $75,518 \div 718$.

9. $40,698 \div 399$.

10. $38,214 \div 386$.

11. $51,171 \div 111$.

12. $10,368 \div 144$.

13. $27,264 \div 96$.

14. $88,534 \div 184$.

15. $20,615 \div 95$.

16. $45,579 \div 209$.   ,

17. $51,867 \div 112$.

18. $309,927 \div 309$.

19. $765,870 \div 98$.

20. $1,536 \div 16$.

21. 2,304 *lbs.* $\div 12$.

22. 1,360 *yds.* $\div 16$ *yds.*

23. 47,708 *ft.* $\div 27$ *ft.*

24. $71,556 \div 201$.

25. 30,056 *lbs.* $\div 884$ *lbs.*

26. 68,541 *mi.* $\div 341$ *mi.*

27. \$14,874 by 402.

28. 14,430 *lbs.* by 74 *lbs.*

29. 300,360 *miles* by 120.

30. 61,712 *horses* by 304.

31. \$722,631 by \$91.

32. \$317,094 by \$82.

33. 1,731,195 *ft.* by 73 *ft.*

34. 7,318,080 by 55.

35. 76,131,702 by 46.

36. 31,231,737 by 37.

37. 13,261,467 by 381.

38. 1,281,524 by 761.

39. 13,189,212 by 937.

40. 728,807 by 731.

41. $\$1,477.35 \div 45 = ?$

42. $947,387 \div 54 = ?$

43. $145,260 \div 108 = ?$

44. $14,420,946 \div 74 = ?$

45. $\$813,204.25 \div 25 = ?$

46. $\$24,411.75 \div 75 = ?$

47. $435,780 \div 216 = ?$

48. $444,312 \div 825 = ?$

49. $\$821.52 \div 63 + \$7.19 \times 8 = ?$

50. $\$5,099.22 \div 57 + \$3.74 \times 156 = ?$

51. $6,574 \div 346 + 12,325 \div 29 = ?$

52. $17,612 \div 518 + 10,323 \div 279 = ?$

53. $(4,320 + 6,003) \div (109 + 170) = ?$

54. $(63,481 - 20,900) \div (119 - 70) = ?$

55. $(\$69.80 - \$12) \div (\$1.15 + \$2.25) = ?$

CONTRACTIONS IN DIVISION.

42. The operation of division may often be shortened when the divisor can be separated into factors. In this case, we divide the dividend by one of the factors; we then divide the quotient by another factor; and so on, till all the factors have been used (Arts. **34, 37**).

EXAMPLES.

1. Divide 19,866 by 77, that is, by 11 × 7.

EXPLANATION.—Here we divide the dividend by 11, and then we divide the resulting quotient by 7. Because each division is exact there is no remainder.

OPERATION.

$$11\ )\ 19866$$
$$7\ )\ 1806$$
$$258\ \textit{Ans.}$$

2. Divide 1,592 by 35, that is, by 7 × 5.

EXPLANATION.— Here we divide in succession by 7 and by 5. The first step shows that 1,592 contains 227 *sevens* and 3 *ones ;* the second step shows that 227 *sevens* contains 45 *thirty fives* and 2 *sevens*.

OPERATION.

$$7\ )\ 1592$$
$$5\ )\ 227\ .\ 3 \quad \text{First remainder.}$$
$$45\ .\ 2 \quad \text{Second remainder.}$$

But 2 *sevens* are equal to 14 *ones ;* hence, the true remainder is 14 + 3 or 17, and the result of the operation is 45$\frac{17}{35}$. Hence, to find the true remainder in all similar cases we *multiply the second remainder by the first divisor, and to the result add the first remainder.*

In like manner to find the true remainder when there are three partial divisions, we *multiply the third remainder by the second divisor, and to the product add the second remainder, we then multiply this result by the first divisor, and to the product add the first remainder.*

3. Divide 7,445 by 84, that is, by 7 × 4 × 3.

EXPLANATION.—Here the several divisors are 7, 4, and 3 ; the corresponding remainders are 4, 3, and 1 ; and the final quotient is 88. The true remainder is (1 × 4 + 3) × 7 + 4, or 53 ; hence, the result of the division is 88$\frac{53}{84}$.

4. 8,154 ÷ 99 = ?   5. 1,913,578 ÷ 42 = ?

6. $15,336 \div 72 = ?$ 9. $1,461,870 \div 7 \times 5 \times 3 = ?$
7. $93,312 \div 108 = ?$ 10. $26,964 \div 11 \times 5 \times 2 = ?$
8. $674,201 \div 110 = ?$ 11. $93,696 \div 11 \times 7 \times 3 = ?$

To divide by 10, 100, 1000, &c., we *point off as many figures from the right of the dividend as there are ciphers in the divisor ; the part on the left is the quotient and the part on the right is the remainder,* (Art. 30).

12. Divide 8,759 by 100.   *Ans.* $87\frac{59}{100}$.

13. 746 by 10.   15. 4,981 by 100.   17. 3,425 by 100.
14. 1,382 by 100.   16. 8,637 by 1000.   18. 94,276 by 10.

NOTE.—If the divisor is composed of a significant part followed by ciphers, we cut off the ciphers and also the same number of figures from the right of the dividend ; we then divide the remaining part of the dividend by the significant part of the divisor ; to find the true remainder we annex to the partial remainder the figures cut off from the dividend.

19. Divide 37,843 by 2,500.   *Ans.* $15\frac{343}{2500}$.

'EXPLANATION.—The operation is equivalent to dividing first by 100 and then by 25. The first partial remainder is 43, the second partial remainder is 3, and the first divisor is 100 ; hence, by the rule, we have the true remainder equal to $3 \times 100 + 43$, or 343.

OPERATION.

$25{,}00\ )\ 378{,}43\ (\ 15$
$\underline{25}$
128
$\underline{125}$
True remainder, 343

20. Divide 98,742 by 1,700.   *Ans.* $58\frac{142}{1700}$.

Perform the following indicated divisions :

21. $8,436 \div 2,100.$ 24. $2,564,310 \div 84,000.$
22. $8,566 \div 2,500.$ 25. $217,896 \div 7,200.$
23. $17,439 \div 1,700.$ 26. $1,310,741 \div 64,000.$

NOTE.—If both dividend and divisor terminate in ciphers, we strike off from the right of each as many as are common to both, and then perform the division.

27. Divide 875,000 by 2,500.

EXPLANATION.—Striking off two ciphers from each is equivalent to dividing both by 100, which obviously does not affect the resulting quotient.

OPERATION.

$25{,}00) 8750{,}00 (350$
75
$\overline{125}$
125

Perform the following divisions:

28. 1,831,200 by 240.
29. $1,350,500 by 3,650.
30. 687,500 *yds.* by 27,500 *yds.*
31. 201,600 by 3,600.
32. 41,580 by 540.
33. 71,820 by 87.
34. 1,749,600 by 360.

35. 98,710 by 8,400.
36. 66,920 by 8,800.
37. 8,623,000 by 250.
38. 47,890 by 8,600.
39. 35,100 by 4,800.
40. $1,400 by $270.
41. 368,000 by 4,200.

PRACTICAL PROBLEMS.

43. The following problems afford exercises in review of the four *fundamental operations,* **Addition, Subtraction, Multiplication,** and **Division.**

1. An estate worth $41,185 was divided equally among 5 persons; what was the share of each ?   *Ans.* $8,237.

2. An estate worth $41,185 was divided equally among a certain number of heirs so that each received $8,237; how many heirs were there ?   *Ans.* 5.

3. The capital of a joint-stock company is $13,125 and is divided into 175 shares; what is the value of each share ?   *Ans.* $75.

4. If a ship sails 5,712 miles in 48 days, how many miles does she sail per day ?   *Ans.* 119.

5. If a ship sails 114 miles in 1 day, how many days will it take her to sail 2,622 miles ?   *Ans.* 23.

6. A farmer paid $13,216 for a farm of 112 acres; how much did he pay per acre ?

7. How many acres of land can be bought with $26,432, at the rate of $59 per acre ?

8. In a field of corn there are 21,033 hills and each row contains 171 hills; how many rows are there ?

9. The mean diameter of the earth is 7,912 miles, and that of the sun is 854,496 miles; how many diameters of the earth are there in the sun's diameter ?

10. A grocer bought 55,664 pounds of flour put up in barrels, each of which contained 196 pounds; how many barrels were there in the lot?

11. There are 4,032 yards of cloth in 96 equal pieces; how many yards are there in each piece ?

12. A field produces 3,404 bushels of oats at the rate of 37 bushels per acre; how many acres are there in the field ?

13. Twenty pieces of cloth contain 39 yards each; 32 pieces contain 38 yards each; and 17 pieces contain 43 yards each; how many yards in all ?

14. A merchant bought 175 yards of cloth at 7 dollars per yard and afterwards sold 72 yards at 9 dollars per yard and the remainder at 8 dollars per yard; how much did he gain ?

15. A dealer bought 27 barrels of flour at $14 per barrel and gave in exchange 32 cords of wood at $8 per cord and paid the balance in cash; how much cash did he pay ?

16. A man's income is $3,150 per year and his expenses are $2,817 per year; how much can he save in 6 years ?

17. A farmer bought 32 acres of land at $95 per acre,

71 acres at $47 per acre, 38 acres at $62 per acre, and 19 acres at $88 per acre ; what did he pay for the whole ?

18. The factors of one number are 19, 17, and 23 ; of another number, 31, 29, and 11 ; and of a third number, 77 and 83 : what is the sum of the numbers ?

19. Two men set out from the same point and travel in opposite directions ; the first travels at the rate of 43 miles per day, and the second at the rate of 37 miles per day : how far apart are they at the end of 7 days ?

20. A farmer bought 6 oxen at $65 each, 12 cows at $42 each, and 142 sheep at $6 each ; what did he pay for the whole ?

21. In a freight car there are 6 boxes of goods, each weighing 382 pounds ; 13 barrels, each weighing 218 pounds ; and 37 bags, each weighing 179 pounds : how many pounds in all ?

22. In 19 bales of cloth, each bale containing 16 pieces, and each piece containing 42 yards, how many yards ?

23. What number multiplied by 86 will give the same product as 163 multiplied by 430 ?

24. How many yards of muslin at 14 cents a yard must be given in exchange for 35 bushels of oats at 56 cents a bushel ?

25. A., B., and C. enter into partnership ; A. puts in $7,200, B. puts in $700 more than A., and C. puts in $550 less than A. and B. together : what is the capital of the firm ?

26. A.'s income is 5 times B.'s, B.'s income is 3 times C.'s, and C.'s income is $1,325 ; what is the entire income of A., B., and C. ?

27. A farmer bought 154 acres of land at $64 per acre, and sold the whole for $11,704; how many dollars did he gain per acre?

28. The distance from New York to Albany is 144 miles, and each mile contains 5,280 feet; how many hours will it take a man to walk from New York to Albany if he walks at the rate of 352 × 60 feet an hour?

29. The sum of two numbers is 10,370, and the second is 4 times the first; what are the numbers?

30. The first of three numbers is 24, the second is 3 times the first, and the third is 4 times the sum of the first and second; what is the difference between the second and third?

31. Write down 4,617, multiply it by 12, divide the product by 9, add 365 to the quotient, and from the sum subtract 5,521; what is the final result?

32. Mrs. White has 3 houses worth $12,530, $11,324, and $9,875, also a farm worth $6,720. To her daughter she gave one third the value of the houses and one fourth the value of the farm, and then she divided the remainder equally between her two sons; how much did each receive?

33. What is the difference between the cost of 425 sheep at $4.75 apiece and 38 cows at $48.25 apiece?

34. The distance from Chicago to San Francisco is 2,448 miles; how long will it take a man to walk the whole distance at the rate of 24 miles a day?

35. Two men had an equal interest in a herd of cattle; one took 72 at $35 apiece and the other took the rest at $42 apiece; how many cattle in the herd?

36. A man bought 4 horses at $116 apiece and 2 colts at $56 apiece, and paid for them in flour at $12 a barrel; how many barrels of flour did it require to make the payment?

37. A man travels due north for 7 days at the rate of 37 miles a day; he then returns on his path at the rate of 29 miles a day; how far is he from the starting point at the end of 12 days travel?

38. A man bought 742 acres of land at $18 an acre; he sold at one time 211 acres at $22 an acre and at another time he sold 184 acres at $25 an acre; at what rate per acre must he sell the rest to gain $3,867?

39. A. bought a farm for $3,612; he sold half of it at $56 an acre and received for it $2,408: how many acres did he buy and what did he give per acre?

40. How many horses worth $112 apiece can be bought for 28 oxen worth $63 each, 52 cows worth $42 each, 175 sheep worth $6 each, and $2,394 in cash?

REVIEW QUESTIONS.

(35.) What is division? Define the dividend; the divisor; the quotient; and the remainder. When is division exact? What are factors? (36.) What does the sign of division indicate when placed between two numbers? In what other ways may division be indicated? (37.) What is the relation between multiplication and division? (38.) State the leading principle of division. (39.) What is short division? What is long division? (40.) Give the rule for short division. Method of Proof? (41.) Give the rule for long division. (42.) How do you divide by a composite number? What is the method of determining the true remainder? How do you divide by 10, 100, 1,000, etc.? How do you divide by a number that ends in ciphers? How do you divide when both dividend and divisor terminate in ciphers? (43.) What are the fundamental operations of arithmetic?

V. FACTORING AND CANCELLING.

DEFINITIONS.

44. A **Factor** of a number is one of its exact integral divisors, (Art. **35**). Thus, 2, 3, and 4, are factors of 12.

45. A **Composite** number is a number composed of two or more **Integral Factors** (Art. **34**). Thus, 15 is a composite number, because it is the product of 3 and 5.

A **Prime** number is one that cannot be separated into any integral factors except 1 and the number itself. Thus, 2, 3, 5, etc., are prime numbers.

46. Factoring is the operation of separating a number into integral factors.

The factors of a number may be either *prime,* or *composite.* Composite factors may themselves be factored, and so on, till all the factors are prime. Thus, $24 = 2 \times 12 = 2 \times 2 \times 6 = 2 \times 2 \times 2 \times 3$; hence, the prime factors of 24 are 2, 2, 2, and 3.

MENTAL EXERCISES.

1. What is the product of 2 and 3 ? What are the factors of 6 ? of 9? of 15 ? of 77 ? of 121 ? of 144 ?

2. What is the continued product of 2, 3, and 7 ? of 2 and 3? of 2 and 7 ? of 3 and 7 ? What are the prime factors of 42 ? What are the composite factors of 42 ? In how many ways may 42 be factored ?

Note.—Because every number is the product of 1 and of the number itself, these numbers are not specially considered in the operation of factoring.

3. What are the prime factors of 4 ? of 9 ? of 27 ? of 81 ? of 121 ? of 143 ? of 187 ? of 198 ? of 225 ?

Note.—The product of two or more factors that are equal is called a **power.** The name of the power depends on the *number* of

3

equal factors. Thus, 3 × 3, or 9, is the *second power*, or the *square* of 3 ; 3 × 3 × 3, or 27, is the *third power*, or the *cube* of 3 ; 3 × 3 × 3 × 3, or 81, is the *fourth power* of 3 ; and so on. Every lower power of a number is a factor of a higher power of the same number.

4. What are the factors of 81 ? of 27 ? of 9 ? How many prime factors has 81 ? How many has 27 ? How many has 9 ? What is the square of 6 ? the third power of 5 ? the fourth power of 4 ? the fifth power of 2 ?

5. What is the second power of 10 ? the third power of 10 ? the fourth power of 10 ? How many ciphers are required to write the square of 10 ? the cube of 10 ?

6. What is the fifth power of 10 ? How many ciphers in the fifth power of 10 ? What power of 10 is 100 ? 1,000 ? 10,000 ? 100,000 ? 1,000,000 ?

PRINCIPLES OF FACTORING.

47. The operation of resolving a number into prime factors depends on the following principles :

1°. *A number is equal to the continued product of all its prime factors.*

2°. *If a number is divided by one of its prime factors, the quotient is equal to the continued product of all the others.*

OPERATION OF FACTORING.

48. Let 210 be separated into prime factors.

EXPLANATION.—We first divide by 2, which is a prime factor ; we next divide the first quotient by 3, which is also a prime factor ; we then divide the second quotient by 5, and find 7 for a quotient. The numbers 2, 3, 5, and 7 are the required factors ; that is,

OPERATION.

$$2)\overline{210}$$
$$3)\overline{105}$$
$$5)\overline{35}$$
$$7$$

$$210 = 2 \times 3 \times 4 \times 5.$$

In like manner other composite numbers may be factored; hence, the following

RULE.

Divide the given number by one of its prime factors; then divide the quotient by one of its prime factors; and so on, till a quotient is found that is a prime number; the several divisors and the last quotient are the required factors.

NOTE.—It will be found convenient to begin the division with the smallest prime factor.

EXAMPLES.

Resolve the following numbers into their prime factors:

1. 42.   *Ans.* $2 \times 3 \times 7$.

NOTE.—If there are more than two factors in any indicated product, the sign of multiplication may be replaced by a simple dot; thus, $2 \cdot 3 \cdot 7$ is equivalent to $2 \times 3 \times 7$.

2. 180. *Ans.* $2 \cdot 2 \cdot 3 \cdot 3 \cdot 5$.   5. 770. *Ans.* $2 \cdot 5 \cdot 7 \cdot 11$.

3. 378. *Ans.* $2 \cdot 3 \cdot 3 \cdot 3 \cdot 7$.   6. 1,575. *Ans.* $3 \cdot 3 \cdot 5 \cdot 5 \cdot 7$.

4. 330. *Ans.* $2 \cdot 3 \cdot 5 \cdot 11$.   7. 3,850. *Ans.* $2 \cdot 5 \cdot 5 \cdot 7 \cdot 11$.

The operation of factoring is principally performed by inspection and trial. It may sometimes be facilitated by using the following

TABLE OF PRIME NUMBERS FROM I TO 150.

1	2	3	5	7	11	13	17	19
23	29	31	37	41	43	47	53	59
61	67	71	73	79	83	89	97	101
103	107	109	113	127	131	137	139	149

Find the prime factors of the following numbers:

8. 402.   10. 3,290.   12. 1,095.

9. 1,659.   11. 1,554.   13. 2,310.

| 14. 2,730. | 16. 7,644. | 18. 786. |
| 15. 17,160. | 17. 1,872. | 19. 3,136. |

NOTE.—If the final digit of a number is 0, 2, 4, 6, or 8, the number is divisible by 2.

If the sum of the digits of a number is divisible by 3, the number itself is divisible by 3.

If the final digit of a number is 5, the number is divisible by 5.

20. 930.	23. 1,738.	26. 1,105.
21. 1,455.	24. 3,255.	27. 3,171.
22. 3,685.	25. 1,001.	28. 2,873.

CANCELLATION.

49. Cancellation is the operation of striking out one or more factors that are common both to the dividend and the divisor of an indicated division.

The operation is performed by drawing a line across the factor that is to be *struck out*, or *cancelled.* Thus, in the expression $\dfrac{\cancel{2}\cdot\cancel{3}\cdot 4}{\cancel{2}\cdot\cancel{3}\cdot 7}$ the factors 2 and 3 are cancelled.

OBJECT AND PRINCIPLES OF CANCELLATION.

50. The operation of division may sometimes be shortened by cancelling factors common to both dividend and divisor. This method depends on the following principles :

1°. *Striking out a factor of a number is equivalent to dividing the number by that factor.*

2°. *If both dividend and divisor are divided by the same number the quotient is not changed.*

MENTAL EXERCISES.

1. What is the quotient of 3 × 8 by 3 × 4? of 8 by 4? What is the effect of cancelling 3 in both dividend and divisor ?

2. Divide $3 \times 3 \times 7$ by $3 \times 3 \times 2$. 7 by 2. 7×7 by 7×2.

APPLICATIONS.

51. Let it be required to divide $2.5.7.11$ by $5.7.7$.

EXPLANATION.—Having indicated the division, we cancel all the factors common to the dividend and divisor; we then divide 2×11, or 22, by 7.

OPERATION.

$$\frac{2.5.7.11}{5.7.7} = \frac{2.\cancel{5}.\cancel{7}.11}{\cancel{5}.\cancel{7}.7} = \frac{22}{7} = 3\tfrac{1}{7}.$$

In like manner we treat all similar cases; hence, the

RULE.

I. Indicate the division, and strike out all the factors that are common to the dividend and divisor.

II. Divide the product of the factors that remain in the dividend by the product of those that remain in the divisor.

EXAMPLES.

Perform the following operations, using the method of cancellation :

1. $2.3.5.7.11 \div 2.3.5.19.$ *Ans.* $\frac{77}{19} = 4\tfrac{1}{19}.$

2. $3.3.3.5.5.7.13 \div 2.2.2.3.3.3.5.5.$

   Ans. $11\tfrac{3}{8}.$

NOTE.—If all the factors in either dividend or divisor are cancelled, the unit 1 will be left; if it is in the dividend it *must* be retained, if it is in the divisor it may be omitted.

4. $2.3.5.7 \div 2.2.5.7.8.$ *Ans.* $\frac{3}{16}.$

5. $3.5.5.7.13 \div 3.5.7.$ *Ans.* 65.

NOTE.—The operation of cancellation should be performed mentally.

6. Divide $14\tfrac{1}{4} \times 56$ by 96. *Ans.* 84.

EXPLANATION.—Having indicated the division, we see that 12 is a factor of 144 and of 96; we therefore mark out 144, replacing it by 12, and 96, replacing it by 8, as shown in the margin. This

OPERATION.

$$\frac{\overset{12}{\cancel{144}} \times 56}{\cancel{96}} = \frac{12 \times \overset{7}{\cancel{56}}}{\cancel{8}} = 84.$$

reduces the operation to dividing 12 × 56 by 8. We now see that 8 is a factor of 56 and also of 8; we therefore cancel 56, replacing it by 7, and 8, dropping the factor 1. The result is 84.

Perform the following indicated operations:

7. 168 × 216 ÷ 42 × 54. 16. 97 × 30 ÷ 37 × 3.

8. 9 × 24 × 31 ÷ 72. 17. 161 × 15 ÷ 161 × 3.

9. 364 × 42 ÷ 14 × 21. 18. 17 × 11 × 4 ÷ 34 × 22.

10. 36 × 37 ÷ 27 × 11. 19. 26 × 33 ÷ 46 × 11.

11. 48 × 125 ÷ 3 × 5. 20. 34 × 26 ÷ 39 × 17.

12. 342 × 6 ÷ 36. 21. 57 × 18 × 4 ÷ 34 × 8.

13. 1,323 × 5 ÷ 63 × 5. 22. 114 × 22 ÷ 11 × 76.

14. 147 × 9 × 5 ÷ 22 × 21. 23. 170 × 55 ÷ 85 × 55.

15. 27 × 200 ÷ 18 × 56. 24. 169 × 7 ÷ 13 × 14.

PRACTICAL PROBLEMS.

1. How many boxes of tea, each containing 24 pounds, at 75 cents a pound, must be given for 145 bags of wheat, each bag containing 2 bushels, at 180 cents a bushel?

Ans. 29 boxes.

SOLUTION. 145 × 2 × 180 ÷ 24 × 75 = 29 *boxes, Ans.*

2. A. worked 18 days at $3 per day, for which he received 6 barrels of flour; how much was the flour worth per barrel? *Ans. $9.*

3. A man buys 3 pieces of cotton cloth, each containing 42 yards, at 13 cents per yard, and pays for it in butter at 21 cents per pound; how many pounds of butter must he give? *Ans. 78 lbs.*

4. Bought 15 barrels of apples, each containing 3 bushels, at 84 cents a bushel; how many cheeses, each weighing 45 pounds, at 12 cents per pound, will pay for the apples ? *Ans. 7 cheeses.*

5. A farmer sold 2 loads of hay each weighing 2,240 *lbs.*, at 1 cent a pound, for which he received 4 pieces of sheeting, each containing 40 *yards;* what did the sheeting cost per yard ?

6. How many bushels of corn at 93 *cts.* a bushel will pay for 2 barrels of sugar each barrel containing 372 *lbs.*, the sugar being worth 8 *cts.* a pound ?

7. A man received for 9 days' work, 3 barrels of flour worth $6 a barrel; what did he receive for each day's work ?

8. How many firkins of butter, each containing 50 *lbs.*, at 18 *cts.* a pound, must be given for 3 *bar.* of sugar, each containing 200 *lbs.*, at 9 *cts.* a pound ?

9. How many boxes of tea, each containing 24 *lbs.*, worth 75 *cts.* a pound, must be given for 4 bins of wheat, each containing 145 bushels, and worth $1.80 a bushel ?

10. How many pounds of butter at 24 *cts.* a pound will buy 18 *yds.* of cotton at 36 *cts.* a yard ?

REVIEW QUESTIONS.

(**44.**) What is a factor of a number ? (**45.**) What is a composite number ? A prime number ? Illustrate. (**46.**) What is factoring ? (**47.**) What are the principles of factoring? (**48.**) What is the rule for resolving a number into prime factors? When there are more than two factors in a product, how may the multiplication be indicated ? (**49.**) What is cancellation ? How is it performed ? (**50.**) What is the object of cancellation ? On what principles does it depend ? (**51.**) Give the rule for shortening division by cancellation.

VI. GREATEST COMMON DIVISOR AND LEAST COMMON MULTIPLE.

DEFINITIONS.

52. A **Common Divisor** of two or more numbers is a number that will exactly divide each of them. Thus, 4 is a common divisor of 8, 16, and 32.

A common divisor of two or more numbers is also called a common measure of those number.

The **Greatest Common Divisor** of two or more numbers is the greatest number that will exactly divide them all. Thus, 8 is the greatest common divisor of 8, 16, and 32.

Numbers that have no common divisor, except 1, are said to be **prime with respect to each other.**

MENTAL EXERCISES.

1. What number will exactly divide both 15 and 20? What is their common divisor?

2. Name all the exact divisions of 30; of 42. What divisors are common to 30 and 42? What is their greatest common divisor?

3. What are the prime factors of 70? of 50? What prime factors are common to 70 and 50? What is their product? What is the greatest common divisor of 70 and 50?

4. What are the prime factors of 18? of 12? of 18—12? What is the greatest common divisor of 18 and 12? of 18 and 18 — 12? of 18 — 12 and 8? What is the greatest common divisor of 121 and 99?

METHODS AND PRINCIPLES.

53. There are two methods of finding the greatest common divisor of two or more numbers: 1°. **By factors**; and 2°. **By continued division.**

Both methods depend on the following principles:

1°. *Any factor common to two or more numbers is a common divisor of those numbers.*

2°. *The greatest common divisor is equal to the continued product of the prime factors that are common to all the numbers.*

METHOD BY FACTORS.

54. Let it be required to find the greatest common divisor of 126, 210, and 546.

Resolving the numbers into prime factors, we have,

$126 = 2.3.3.7,\ 210 = 2.3.5.7,$ and $546 = 2.3.7.13.$

The factors 2, 3, and 7 are *common* to all the numbers, and they are the only ones that are common; hence, their product is the greatest common divisor of the given numbers; denoting the greatest common divisor by the initials *g. c. d.*, we have *g. c. d.* $= 42.$

All similar cases may be treated in like manner; hence, the following

RULE.

I. Resolve the numbers into prime factors.

II. Find the continued product of the prime factors common to all the numbers.

EXAMPLES.

1. What is the greatest common divisor of 168, 216, and 408, that is, of $2.2.2.3.7,\ 2.2.2.3.9,$ and $2.2.2.3.17$?   *Ans.* $2.2.2.3 = 24.$

Find the greatest common divisors of the following groups of numbers:

2. 408, and 740.

3. 90, 315, and 810.

4. 441, and 567.

5. 195, 285, and 315.

6. 462, 726, and 1,254.

7. 1,470, 2,310, and 2,730.

8. 320, 1,216, and 6,400.

9. 540, 648, and 756.

10. 567, 648, and 729.

11. 84, 126, and 210.

12. 4,410, and 3,150.

13. 1,335, and 1,869.

14. 1,584, and 1,188.

15. 26,195, and 273.

ADDITIONAL PRINCIPLE.

55. The greatest common divisor of two numbers will divide their remainder after division.

For, let 8, which is the greatest common divisor of 88 and 24, be taken as a *unit:* the two numbers, expressed in terms of this unit, are 11 *eights* and 3 *eights*. Let the greater number be divided by the less.

EXPLANATION. — We see that 11 *eights* contains 3 *eights*, 3 times with a remainder equal to 2 *eights*.

OPERATION.

Divisor. Dividend.

3 *eights*)11 *eights*(3 ... Quotient.

9 *eights*

2 *eights* ... Remainder.

Now, 11 *eights*, 3 *eights*, and 2 *eights* are all divisible by 8, and the quotients are prime with respect to each other ; hence, 8 is the greatest common divisor of 3 *eights* and 2 *eights*, as well as of 11 *eights* and 3 *eights*, that is, the greatest common divisor of the given numbers is also the greatest common divisor of the less number and of their remainder after division : hence, the following

PRINCIPLE.

The greatest common divisor of two numbers is the same as the greatest common divisor of the smaller number and of their remainder after division.

METHOD BY CONTINUED DIVISION.

56. Let it be required to find the greatest common divisor of 88 and 24.

EXPLANATION.—Dividing 88 by 24, we find 16 for a remainder; then dividing 24 by 16, we find 8 for a remainder; then dividing 16 by 8, we find 0 for a remainder. Hence, 8 is the greatest common divisor of 16 and 24, (Art. **55**); it is therefore the greatest common divisor of 24 and 88.

In like manner, all similar cases may be treated; hence, the following

OPERATION.

$$24)88(3$$
$$\underline{72}$$
$$16)24(1$$
$$\underline{16}$$
$$8)16(2$$
$$\underline{16}$$
$$0$$

RULE.

I. Divide the greater number by the less and find the remainder.

II. Take the divisor for a new dividend and the remainder for a divisor, and proceed as before.

III. Continue the operation till a remainder is found that will exactly divide the preceding divisor; this will be the greatest common divisor of the given numbers.

EXAMPLES.

Find the greatest common divisors of the following groups of numbers:

1. 3,471 and 2,136.
2. 1,584 and 3,168.
3. 2,898 and 7,866.
4. 3,724 and 5,852.
5. 3,444 and 2,268.
6. 10,395 and 16,797.
7. 667 and 391.
8. 10,353 and 14,877.
9. 4,410 and 5,670.
10. 3,471 and 1,869.
11. 1,584 and 2,772.
12. 432 and 1,224.
13. 945 and 3,240.
14. 1,080 and 1,224.

To find the greatest common divisor of more than two numbers, begin with the least and find the greatest common divisor of two, then of that result and the third, then of that and the fourth, and so on to the last.

15. 805, 1,311, and 1,978.	21. 740, 999, and 1,147.
16. 504, 5,292, and 1,512.	22. 108, 216, and 324.
17. 837, 1,134, and 1,347.	23. 803, 949, and 1,241.
18. 492, 744, and 1,044.	24. 836, 988, and 1,444.
19. 944, 1,488, and 2,088.	25. 935, 1,045, and 1,265.
20. 216, 408, and 740.	26. 581, 1,079, and 913.

LEAST COMMON MULTIPLE.

DEFINITIONS.

57. A **Multiple** of a number is a number that is exactly divisible by it. Thus, 12 is a multiple of 6.

A **Common Multiple** of two or more numbers is a number that is exactly divisible by each. Thus, 48 is a common multiple of 4, 6, and 8.

The **Least Common Multiple** of two or more numbers is the least number that is exactly divisible by each. Thus, 24 is the least common multiple of 4, 6, and 8.

MENTAL EXERCISES.

1. What are the prime factors of 4? of 6? of 12? What is the least number that can be divided by both 4 and 6? What is the least common multiple of 4 and 6? of 12 and 30? of 18 and 48?

2. How many times is 2 a factor of 18? of 60? of 180? How many times is 3 a factor of 18? of 60? of 180? Are there any prime factors in 18, or in 60, not in 180? What is the least common multiple of 18 and 60?

PRINCIPLES.

58. The operation of finding the least common multiple of two or more numbers depends on the following *principles:*

1°. *The least common multiple must contain every prime factor of each number.*

2°. *It must contain every prime factor the greatest number of times it enters any of the numbers.*

OPERATION OF FINDING THE LEAST COMMON MULTIPLE.

59. Let it be required to find the least common multiple of 12, 25, and 90.

Resolving the given numbers into prime factors, we have,

$$12 = 2.2.3, \quad 25 = 5.5, \quad \text{and} \quad 90 = 2.3.3.5.$$

The least common multiple must contain the factor 2 *twice* to be divisible by 12, it must contain the factor 3 *twice* to be divisible by 90, and it must contain the factor 5 *twice* to be divisible by 25. Hence, if we denote the least common multiple by the initials *l. c. m.*, we have,

$$l. c. m. = 2.2.3.3.5.5 = 900.$$

The practical method of factoring is as follows:

EXPLANATION.—Having written the numbers in a line, we see by inspection that 2 is a prime factor of 12 and 90. We therefore write 2 on the left as a divisor. Dividing 12 and 90 by 2, we write the quotients, and also the undivided number, 25, in the second line. We then see that 5 is a prime factor

OPERATION.

2	12,	25,	90
5	6,	25,	45
3	6,	5,	9
	2,	5,	3

of 25 and 45; writing it on the left and proceeding as before, we find the numbers in the third line. We then see that 3 is a prime factor of 6 and 9; proceeding as before, we find the numbers in the fourth line, all of which are prime with respect to each other. Here, we have resolved 12 into the factors 2, 3, and 2 ; 25 into the factors 5, and 5 ; and 90 into the factors 2, 5, 3, and 3. Hence, from the principles of Art. 58, we have $l. c. m. = 2 . 2 . 3 . 3 . 5 . 5 = 900$.

In like manner we can find the least common multiple of any other group of numbers ; hence, the following

RULE.

I. Write the numbers in a line and divide by any prime factor that is contained in two or more, writing the quotients and the undivided numbers in the line below.

II. Then operate on the second line of numbers in the same manner, and so on, till a line of numbers is found that are prime with respect to each other.

III. Find the continued product of the numbers in the last line and of the divisors used; this will be the least common multiple of the given numbers.

EXAMPLES.

Find the least common multiple of the following groups:

1. 3, 4, 8, and 12.	11. 84, 100, and 224.
2. 6, 7, 8, 9, and 10.	12. 49, 56, 63, and 84.
3. 4, 6, 9, 14, and 16.	13. 20, 126, 150, and 490.
4. 12, 48, 18, and 70.	14. 84, 150, and 1,225.
5. 14, 20, 198, and 210.	15. 39, 52, 78, and 117.
6. 8, 18, 20, and 70.	16. 130, 390, and 338.
7. 9, 18, 27, 36, 54, 45.	17. 136, 170, and 425.
8. 7, 15, 21, 28, and 35.	18. 171, 285, and 475.
9. 15, 16, 18, 20, and 24.	19. 275, 385, and 539.
10. 49, 14, 84, 168, and 98.	20. 126, 189, and 56.

PRACTICAL PROBLEMS.

1. What is the *g. c. d.* of $18 and $45 ?

2. What is the *l. c. m.* of 12 *ft.*, and 90 *ft.* ?

3. A farmer has 225 *bu.* of oats, 135 *bu.* of wheat, and 90 *bu.* of rye, which he wishes to put in bins of equal size; each bin to be as large as possible; how many bushels must each hold that all may be filled without mixing the different kinds of grain ?

4. What is the shortest piece of wire that can be cut up into exact lengths of either 6 *ft.*, 8 *ft.*, or 10 *ft.* ?

5. There are three companies of soldiers containing respectively 36, 60, and 84 men, each of which is to be divided into platoons; how many men must be put in a platoon, so that all the platoons shall be equal and each contain the greatest possible number of men ?

6. How many quarts are there in the smallest cask of cider that can be exactly measured by either a 3 *quart*, a 5 *quart*, or a 6 *quart* measure ?

REVIEW· QUESTIONS.

(**52.**) What is a common divisor of two or more numbers? What is the greatest common divisor of two or more numbers? When are numbers prime with respect to each other? (**53.**) What general principles are used in finding the greatest common divisor? (**54.**) Give the rule for finding the greatest common divisor by the method of factors. (**55.**) What additional principle is used? (**56.**) Give the rule for finding the greatest common divisor by the method of continued division. How do you find the greatest common divisor of more than two numbers? (**57.**) What is a multiple of a number? What is a common multiple of two or more numbers? What is their least common multiple? (**58.**) Give the principles used in finding the least common multiple. (**59.**) Give the rule for finding the least common multiple.

FRACTIONS.

I. COMMON FRACTIONS.

DEFINITIONS.

60. If a unit is divided into equal parts, each part is called a **fractional unit**.

If the unit is divided into two equal parts, each is called **one-half**; if into three, each is called **one-third**; if into four, each is called **one-fourth**; and so on.

Fractional units may be written and read as shown below:

$\frac{1}{2}$, $\frac{1}{3}$, $\frac{1}{4}$, $\frac{1}{5}$, $\frac{1}{6}$, $\frac{1}{7}$,   etc.

one-half,   one-third,   one-fourth,   one-fifth,   one-sixth. one-seventh,   etc.

The **Reciprocal of a Number** is 1 divided by that number. Thus, $\frac{1}{2}$ is the reciprocal of 2, $\frac{1}{3}$ is the reciprocal of 3, and so on.

MENTAL EXERCISES.

1. If a unit is divided into 5 equal parts, what is each part called? If it is divided into 9 equal parts, what is each part called? If into 10? If into 12? If into 13?

2. How many *halves* of an apple are there in 1 apple? How many *fifths?* How many *ninths?* How many *tenths?* How many *twelfths? Fifteenths? Twentieths?*

3. What is the reciprocal of 10 ? Of what number is $\frac{1}{10}$ the reciprocal? $\frac{1}{12}$? $\frac{1}{17}$? $\frac{1}{20}$? $\frac{1}{25}$?

61. A **Fraction** is a fractional unit, or a collection of fractional units; thus, *one-half*, *two-thirds*, *four-ninths*, etc., are fractions.

Fractions may be written and read as shown below :

$\frac{3}{4}$, $\frac{2}{9}$, $\frac{5}{7}$, $\frac{8}{11}$,   etc.
three-fourths,   two-ninths,   five-sevenths,   eight-elevenths,   etc.

Fractions written in this manner are called **vulgar,** or **Common Fractions.**

Common fractions are expressed by two numbers, one written above the other, with a line between them. The number below the line is called the **Denominator,** and the one above it is called the **Numerator.** Both *numerator* and *denominator* are called **Terms** of the fraction.

The *denominator* indicates the number of equal parts into which the unit is divided, and the *numerator* shows how many of these parts are taken. Thus, in the fraction $\frac{3}{4}$, the denominator indicates that 1 is divided into 4 equal parts, and the numerator shows that 3 of these are taken.

MENTAL EXERCISES.

1. If 1 is divided into 7 equal parts, what is 1 part called ? What are 4 parts called ? If 1 is divided into 11 equal parts, what are 5 of them called ? 6 of them ?

2. If 1 *yard* is divided into 8 equal parts, what is 1 part called ? 3 parts ? 5 parts ? 7 parts ? How many *eighths of an apple* are there in 1 *apple ?* in 3 *apples ?* in 7 *apples ?* How many *elevenths* in 6 ? in 11 ? in 13 ? in 17 ? in 19 ? How many *twelfths* in 5 ? in 11 ? in 13 ? in 15 ? in 21 ?

3. In the fraction $\frac{5}{8}$, what is the *denominator?* What is the *numerator?* What is the *fractional unit?* How many times is this unit taken ? How many times $\frac{1}{8}$ are $\frac{5}{8}$?

Write the following fractions:

    1. Seven-eighths.   4. Eleven-hundredths.

    2. Four-tenths.   5. Thirteen-twenty fifths.

    3. Nine-twentieths.   6. Sixty-thousandths.

Read the following fractions:

    1. $\frac{3}{4}$, $\frac{10}{11}$, $\frac{8}{19}$, $\frac{7}{25}$, $\frac{16}{38}$, $\frac{27}{100}$, $\frac{125}{1000}$, $\frac{83}{88}$, $\frac{17}{117}$, $\frac{28}{47}$, $\frac{15}{19}$, $\frac{19}{37}$.

    2. $\frac{4}{5}$, $\frac{11}{27}$, $\frac{124}{250}$, $\frac{78}{1000}$, $\frac{92}{185}$, $\frac{412}{655}$, $\frac{22}{87}$, $\frac{115}{225}$, $\frac{77}{1000}$, $\frac{188}{880}$, $\frac{112}{294}$.

62. A fraction is equal to its **numerator divided by its denominator.** Thus, $\frac{3}{4}$ is equal to $3 \div 4$; for, if each of the units in 3 is divided into four equal parts, we shall have 12 such parts, each equal to $\frac{1}{4}$, that is, we shall have 12 *fourths ;* but 12 *fourths* divided by 4 is equal to 3 *fourths*, or to $\frac{3}{4}$, that is, $3 \div 4$ is equal to $\frac{3}{4}$.

63. A fraction may be regarded either as a number, or as an indicated division:

1°. Regarded as a number, the unit is *fractional* and equal to the reciprocal of the denominator. Thus, $\frac{3}{4}$ is a collection of 3 units, each equal to $\frac{1}{4}$, that is, $\frac{3}{4} = 3 \times \frac{1}{4}$.

2°. Regarded as an indicated division, the numerator is the dividend and the denominator is the divisor. Thus, $\frac{3}{4} = 3 \div 4$.

MENTAL EXERCISES.

1. In the fraction $\$\frac{7}{11}$, what is the *fractional unit?* How many times is it taken ? What is 7 times $\$\frac{1}{11}$?

2. Is there any difference between $\frac{9}{10}$ of 1 *pound* and $\frac{1}{10}$ of 9 *pounds?* $\frac{4}{4}$ of 1 dollar, is what part of 4 dollars ? How does $\frac{4}{4}$ of 1 *dollar* differ from $\frac{1}{4}$ of 4 *dollars?*

DEFINITIONS.

64. A **Proper Fraction** is one in which the numerator is less than the denominator; as, $\frac{3}{4}$, $\frac{5}{6}$.

An **Improper Fraction** is one in which the numerator is equal to, or greater than the denominator; as, $\frac{4}{4}$, $\frac{8}{5}$.

NOTE.—If the numerator is less than the denominator, the fraction is less than 1; if the numerator is equal to the denominator, the fraction is equal to 1; if the numerator is greater than the denominator, the fraction is greater than 1.

A **Simple Fraction** is one, both of whose terms are whole numbers; as, $\frac{4}{7}$, $\frac{5}{6}$.

A **Mixed Number** is a number composed of an integral and of a fractional part; as, $2\frac{1}{2}$, $3\frac{1}{4}$.

A **Complex Fraction** is one that has at least one of its terms fractional; as, $\dfrac{\frac{4}{5}}{5}$, $\dfrac{3}{14\frac{1}{2}}$, $\dfrac{15}{(\frac{1}{4})}$, $\dfrac{7\frac{1}{2}}{8\frac{1}{4}}$.

NOTE.—A whole number may be regarded as a fraction whose denominator is 1. Thus, $8 = \frac{8}{1}$.

FUNDAMENTAL PRINCIPLES.

65. Because the numerator shows how many times the fractional unit is taken, we have the following principles:

1°. *Multiplying the numerator of a fraction by any number is equivalent to multiplying the fraction by that number.*

2°. *Dividing the numerator of a fraction by any number is equivalent to dividing the fraction by that number.*

Because the denominator shows the number of equal parts into which we divide the unit 1 to obtain the fractional unit, we have the following principles:

. 3°. *Multiplying the denominator of a fraction by any*

number is equivalent to dividing the fraction by that number.

4°. *Dividing the denominator of a fraction by any number is equivalent to multiplying the fraction by that number.*

From principles 1° and 3°, and from principles 2° and 4°, we have the following principles:

5°. *Multiplying both terms of a fraction by the same number does not alter its value.*

6°. *Dividing both terms of a fraction by the same number does not alter its value.*

REDUCTION OF FRACTIONS.

66. Reduction is the operation of changing the form of a number without altering its value.

The methods of reducing fractions depend on the principles just deduced.

67. To reduce a whole number to the form of a fraction having a given fractional unit.

MENTAL EXERCISES.

1. In 4 *apples*, how many *tenths of an apple?*

EXPLANATION.—In 1 *apple* there are 10 *tenths*, hence in 4 *apples* there are 4 times 10, or 40 *tenths*.

2. How many *quarters of a dollar* must I pay for a hat worth $4 ? How many quarters in 4 ? in 5 ?

3. How many sevenths are there in 12 ? in 15 ? in 17 ? in 30 ? What is the difference between 12 and $\frac{84}{4}$? between 15 and $\frac{104}{7}$? What is the fractional unit of $\frac{104}{7}$?

4. How many *tenths* in 9 ? in 17 ? How many *elevenths* in 11 ? How many in 13 ? in 17 ? in 19 ?

Let it be required to change 17 to the form of a fraction whose unit is $\frac{1}{6}$:

EXPLANATION.—Having written 17 under the form of a fraction, (Art. **64**), we multiply both of its terms by 6, (Prin. 5°), which gives $\frac{102}{6}$.

OPERATION.

$$17 = \frac{17}{1} = \frac{17 \times 6}{1 \times 6} = \frac{102}{6}.$$

In like manner we may treat all similar cases ; hence, the following

RULE.

Multiply the number by the denominator of the given unit and write the product over that denominator.

EXAMPLES.

Reduce

1. 12 to the unit $\frac{1}{4}$.
2. 14 to the unit $\frac{1}{8}$.
3. 7 to the unit $\frac{1}{8}$.
4. 19 to the unit $\frac{1}{4}$.
5. 42 to the unit $\frac{1}{11}$.

Reduce

6. 59 to the unit $\frac{1}{85}$.
7. 212 to the unit $\frac{1}{29}$.
8. 524 to the unit $\frac{1}{17}$.
9. 326 to the unit $\frac{1}{84}$.
10. 426 to the unit $\frac{1}{97}$.

68. To reduce a mixed number to the form of a simple fraction.

MENTAL EXERCISES.

1. In $3\frac{3}{4}$ *pounds*, how many *fourths of a pound ?*

EXPLANATION.—In 3 *pounds* there are 3 × 4, or 12 *fourths of a pound*, hence in $4\frac{3}{4}$ pounds there are 12 + 3 or 15 *fourths of a pound*.

2. How many *tenths of a dollar* must I give for a book worth \$$3\frac{3}{10}$? How many *tenths* in $5\frac{7}{10}$? in $9\frac{9}{10}$? in $11\frac{4}{10}$? How many *twelfths* in $7\frac{3}{12}$? in $13\frac{5}{12}$?

3. How many *ninths* are there in $5\frac{8}{9}$? What is the difference between $5\frac{8}{9}$ and $\frac{47}{9}$? What is the fractional unit of $\frac{42}{9}$? How many *fifteenths* in 7? in 13?

Let it be required to reduce $12\frac{3}{7}$ to the form of a simple fraction:

EXPLANATION.—The mixed number $12\frac{3}{7}$ is equal to $12 + \frac{3}{7}$; reducing its integral part to the unit $\frac{1}{7}$, it becomes $\frac{84}{7}$, that is, the given number is equal to $\frac{1}{7}$, taken $84 + 3$ times, or to $\frac{87}{7}$.

OPERATION.

$$12 + \frac{3}{7} = \frac{84}{7} + \frac{3}{7} = \frac{87}{7}.$$

In like manner we may treat all similar cases; hence, the following

R U L E.

Multiply the entire part by the denominator of the fraction and to the product add the numerator; then place the sum over the given denominator.

E X A M P L E S.

Reduce the following numbers to simple fractions:

1. $3\frac{2}{23}$.
2. $9\frac{7}{8}$.
3. $10\frac{4}{5}$.
4. $12\frac{3}{17}$.
5. $15\frac{2}{23}$.
6. $6\frac{14}{17}$.
7. $101\frac{3}{4}$.
8. $64\frac{3}{5}$.
9. $15\frac{7}{16}$.
10. $12\frac{7}{12}$.
11. $\$19\frac{3}{10}$.
12. $16\frac{1}{4}$ *lbs.*
13. $43\frac{1}{4}$ *yds.*
14. $87\frac{2}{3}$ *hrs.*

15. $27\frac{2}{13}$.
16. $15\frac{3}{17}$.
17. $44\frac{9}{55}$.
18. $31\frac{9}{25}$.
19. $22\frac{7}{16}$.
20. $100\frac{7}{10}$.
21. $102\frac{9}{10}$.
22. $25\frac{3}{11}$.
23. $31\frac{7}{8}$.
24. $118\frac{11}{13}$.
25. $\$119\frac{7}{4}$.
26. $\$316\frac{2}{3}$.
27. $\$177\frac{4}{11}$.
28. $210\frac{3}{4}$ *lbs.*

29. $152\frac{11}{12}$.
30. $207\frac{11}{25}$.
31. $391\frac{11}{14}$.
32. $237\frac{19}{104}$.
33. $215\frac{11}{12}$.
34. $187\frac{11}{33}$.
35. $1,630\frac{2}{13}$.
36. $579\frac{11}{392}$.
37. $4,311\frac{15}{16}$.
38. $3,204\frac{25}{36}$.
39. $910\frac{3}{4}$ *lbs.*
40. $472\frac{4}{11}$ *yds.*
41. $365\frac{1}{4}$ *da.*
42. $\$290\frac{4}{12}$.

69. To reduce an improper fraction to the form of an integral or a mixed number.

MENTAL EXERCISES.

1. In 17 *fifths of a pound,* how many *pounds* and parts of a pound ?

EXPLANATION.—Because 5 *fifths* of a pound make 1 *pound,* 17 *fifths* will make $\frac{17}{5}$, or $3\frac{2}{5}$ *pounds.*

2. In $\frac{45}{10}$ of a *dollar,* how many *dollars* and how many *tenths of a dollar ?* In $\frac{43}{4}$ of a *yard,* how many *yards* and how many *fourths of a yard ?* How many whole feet and what fraction of a foot are there in $\frac{141}{11}$ of a foot ?

3. A boy bought 18 *melons* at $\$\frac{1}{4}$ a piece; what did he pay for them ? What would they have cost at $\$\frac{1}{3}$ each ? at $\$\frac{1}{8}$? at $\$\frac{1}{6}$? at $\$\frac{1}{7}$? at $\$\frac{2}{5}$? at $\$\frac{3}{17}$? at $\$\frac{2}{15}$?

Let it be required to reduce $\frac{104}{22}$ to a mixed number.

EXPLANATION.—We perform the indicated division, that is, we divide the numerator by the denominator and find the quotient, $4\frac{16}{22}$, which is a mixed number.

In like manner we may treat all similar cases; hence, the following

OPERATION.

$22)104(4\frac{16}{22}$

$\underline{88}$

16 Rem.

RULE.

Divide the numerator by the denominator; the quotient will be the required number.

NOTE.—If the remainder is 0, the division is exact and the given fraction is a whole number under a fractional form.

EXAMPLES.

Reduce the following fractions to mixed numbers:

1. $\frac{17}{6}$.

2. $\frac{113}{9}$.

3. $\frac{443}{12}$.

4. $\frac{183}{55}$.

5. $\frac{751}{24}$.

6. $\frac{580}{63}$.

7. $\frac{876}{157}$.

8. $\frac{642}{108}$.

9. $\frac{1000}{111}$.

10. $\frac{3175}{163}$.

11. $\frac{100}{27}$.

12. $\frac{775}{31}$.

13. $\frac{444}{133}$.   19. $\frac{27324}{812}$.   25. $\frac{27342}{188}$.

14. $\frac{266}{64}$.   20. $\frac{85763}{218}$.   26. $\frac{23573}{333}$.

15. $\frac{664}{29}$.   21. $\frac{4136}{38}$.   27. $\frac{43400}{100}$.

16. $\frac{455}{76}$.   22. $\frac{3398}{179}$.   28. $\frac{108700}{27000}$.

17. $\frac{1261}{366}$.   23. $\frac{14235}{812}$.   29. $\frac{84200}{9700}$.

18. $\frac{2565}{604}$.   24. $\frac{48434}{218}$.   30. $\frac{59840}{680}$.

70. To reduce a fraction to its lowest terms.

A fraction is said to be in its **lowest terms** when its terms are *prime* with respect to each other, that is, when they have no common factor.

MENTAL EXERCISES.

1. If $\frac{1}{3}$ is divided into 3 equal parts, what is the value of each part? How many *thirds* are there in 3 *ninths?* in 6 *ninths?* How does $\frac{3}{3}$ of an *apple* differ from $\frac{1}{3}$ of an *apple?* Is there any difference between the values $\frac{6}{9}$ and $\frac{2}{3}$? of $\frac{6}{15}$ and $\frac{2}{5}$? of $\frac{18}{24}$ and $\frac{3}{4}$?

2. If we divide the dividend and divisor by the same number, what effect will it have on the quotient? If we divide both terms of a fraction by the same number, will it affect the value of the fraction?

Let it be required to reduce the fraction $\frac{30}{105}$ to its lowest terms.

EXPLANATION.—We first resolve the terms of the fraction into prime factors, and then strike out those that are common to both, (Principle 6°,

OPERATION.
$$\frac{30}{105} = \frac{2 \cdot 3 \cdot 5}{3 \cdot 5 \cdot 7} = \frac{2}{7}$$

Art. 65); the resulting fraction is equivalent to the given one, and is in its lowest terms.

Since all similar cases may be treated in like manner, we have the following

R U L E .

Resolve the terms of the fraction into prime factors, and cancel all that are common to both.

NOTE.—If the terms cannot be factored by inspection, find their greatest common divisor and divide them both by it.

E X A M P L E S .

Reduce the following fractions to their lowest terms:

1. $\frac{54}{81}$.	14. $\frac{78}{198}$.	27. $\frac{1302}{1785}$.
2. $\frac{63}{135}$.	15. $\frac{66}{792}$.	28. $\frac{504}{1848}$.
3. $\frac{248}{312}$.	16. $\frac{72}{1812}$.	29. $\frac{5184}{6912}$.
4. $\frac{8465}{7335}$.	17. $\frac{284}{403}$.	30. $\frac{3148}{8148}$.
5. $\frac{195}{225}$.	18. $\frac{154}{217}$.	31. $\frac{1888}{1998}$.
6. $\frac{33}{87}$.	19. $\frac{96}{144}$.	32. $\frac{765}{1377}$.
7. $\frac{451}{627}$.	20. $\frac{544}{816}$.	33. $\frac{3648}{3876}$.
8. $\frac{91}{119}$.	21. $\frac{845}{875}$.	34. $\frac{3487}{3383}$.
9. $\frac{408}{408}$.	22. $\frac{1872}{2016}$.	35. $\frac{8148}{8148}$.
10. $\frac{115}{253}$.	23. $\frac{3088}{3188}$.	36. $\frac{8058}{10434}$.
11. $\frac{131}{306}$.	24. $\frac{383}{608}$.	37. $\frac{5484}{8324}$.
12. $\frac{345}{564}$.	25. $\frac{1088}{1875}$.	38. $\frac{4458}{7700}$.
13. $\frac{189}{567}$.	26. $\frac{350}{1015}$.	39. $\frac{1828}{2244}$.

71. To reduce a fraction to an equivalent fraction whose denominator is a multiple of the given denominator.

Let it be required to reduce $\frac{3}{4}$ to an equivalent fraction when the denominator is 12 :

EXPLANATION.—Here we have multiplied both terms of the given fraction by $\frac{12}{4}$, or 3 ; the resulting fraction, $\frac{9}{12}$, is equivalent to the given fraction (Prin. 5°), and its denominator is 12, that is, its fractional unit is $\frac{1}{12}$.

OPERATION.

$$\frac{3}{4} = \frac{3 \times 3}{4 \times 3} = \frac{9}{12}$$

In like manner other fractions may be reduced ; hence, the

R U L E .

Multiply both terms of the fraction by the quotient of the required denominator divided by the given denominator.

E X A M P L E S .

1. Reduce $\frac{2}{5}$ to a fraction whose unit is $\frac{1}{30}$.

Reduce, $Ans.$ $\dfrac{2 \times 6}{5 \times 6} = \dfrac{12}{30}$.

2. $\frac{7}{11}$ to the unit $\frac{1}{44}$.
3. $\frac{3}{13}$ to the unit $\frac{1}{65}$.
4. $\frac{9}{16}$ to the unit $\frac{1}{160}$.
5. $\frac{4}{15}$ to the unit $\frac{1}{120}$.

6. $\frac{3}{11}$ to the unit $\frac{1}{99}$.
7. $\frac{7}{13}$ to the unit $\frac{1}{117}$.
8. $\frac{2}{17}$ to the unit $\frac{1}{153}$.
9. $\frac{4}{15}$ to the unit $\frac{1}{180}$.

72. To reduce two or more fractions to equivalent fractions having a common denominator.

Let it be required to reduce $\frac{2}{3}, \frac{3}{4}$, and $\frac{2}{7}$ to a common denominator :

OPERATION.

EXPLANATION.—Here we have multiplied both terms of each fraction by the product of the denominators of the other fractions ; the denominators of each of the resulting fractions is then equal to the continued product of the denominators of all the fractions.

$$\frac{2}{3} = \frac{2 \times 4 \times 7}{3 \times 4 \times 7} = \frac{56}{84},$$

$$\frac{3}{4} = \frac{3 \times 3 \times 7}{4 \times 3 \times 7} = \frac{63}{84},$$

$$\frac{2}{7} = \frac{2 \times 3 \times 4}{7 \times 3 \times 4} = \frac{24}{84},$$

In like manner we may treat all other groups of fractions hence, the following

R U L E .

Multiply both terms of each fraction by the product of the denominators of all the other fractions.

NOTE.—If the denominators are small, the operation can be performed mentally.

EXAMPLES.

Reduce to a common denominator:

1. $\frac{1}{2}$, $\frac{2}{3}$, and $\frac{3}{4}$.

2. $\frac{2}{3}$, $\frac{3}{4}$, and $\frac{5}{6}$.

3. $\frac{2}{3}$, $\frac{4}{7}$, and $\frac{5}{6}$.

4. $\frac{3}{13}$, $\frac{4}{11}$, and $\frac{7}{16}$.

5. $\frac{23}{36}$, $\frac{11}{18}$, and $\frac{7}{16}$.

6. $\frac{4}{15}$, $\frac{13}{17}$, and $\frac{16}{23}$.

7. $\frac{19}{14}$, $\frac{15}{26}$, and $\frac{7}{68}$.

8. $\frac{4}{11}$, $\frac{11}{13}$, and $\frac{37}{44}$.

9. $\frac{5}{6}$, $\frac{3}{8}$, and $\frac{11}{13}$.

10. $\frac{3}{8}$, $\frac{1}{23}$, and $\frac{29}{44}$.

73. To reduce two or more fractions to their least common denominator.

Let it be required to reduce $\frac{8}{12}$, $\frac{15}{20}$, and $\frac{20}{48}$ to equivalent fractions having the least possible common denominator.

EXPLANATION.—Here we first reduce each fraction to its lowest terms (Art. 70); we next find the least common multiple, 24, of the new denominators, which will be the required denominator; we then divide this multiple by each denominator separately, and multiply both terms of the corresponding fraction by the quotient. Thus, we multiply both terms of $\frac{2}{3}$ by 8, both terms of $\frac{3}{4}$ by 6, and both terms of $\frac{5}{12}$ by 2.

OPERATION.

$$\frac{8}{12} = \frac{2}{3} = \frac{2 \times 8}{3 \times 8} = \frac{16}{24},$$

$$\frac{15}{20} = \frac{3}{4} = \frac{3 \times 6}{4 \times 6} = \frac{18}{24},$$

$$\frac{20}{48} = \frac{5}{12} = \frac{5 \times 2}{12 \times 2} = \frac{10}{24}.$$

In like manner we may treat all similar cases; hence, the following

RULE.

I. Reduce each fraction to its simplest terms.

II. Find the least common multiple of all the denominators for a common denominator; divide this by each denominator separately and multiply the corresponding numerator by the quotient.

EXAMPLES.

Reduce each of the following groups of fractions to a common denominator:

1. $\frac{1}{2}$, $\frac{2}{3}$, $\frac{3}{4}$, and $\frac{5}{12}$.   *Ans.* $\frac{12}{24}$, $\frac{16}{24}$, $\frac{21}{24}$, and $\frac{10}{24}$.

2. $\frac{5}{8}$, $\frac{2}{3}$, $\frac{3}{4}$, and $\frac{4}{15}$.   *Ans.* $\frac{75}{120}$, $\frac{48}{120}$, $\frac{90}{120}$, and $\frac{32}{120}$.

3. $\frac{2}{7}$, $\frac{5}{14}$, $1\frac{8}{21}$, and $\frac{6}{28}$.   17. $\frac{142}{26}$, $\frac{28}{30}$, and $1\frac{5}{7}$.

4. $\frac{7}{7}$, $\frac{5}{11}$, $1\frac{5}{8}$, and $\frac{6}{22}$.   18. $\frac{111}{11}$, $\frac{37}{321}$, and $\frac{19}{428}$.

5. $\frac{3}{11}$, $\frac{6}{26}$, $1\frac{5}{33}$, and $\frac{3}{39}$.   19. $\frac{3}{8}$, $\frac{4}{5}$, $\frac{5}{11}$, and $4\frac{7}{8}$.

6. $\frac{3}{8}$, $1\frac{5}{18}$, and $2\frac{8}{36}$.   20. $1\frac{8}{13}$, $\frac{9}{15}$, $1\frac{4}{4}$, and $\frac{118}{260}$.

7. $2\frac{5}{21}$, $1\frac{8}{16}$, and $1\frac{1}{14}$.   21. $\frac{1}{2}$, $\frac{2}{3}$, $\frac{3}{4}$, $\frac{4}{5}$, and $\frac{5}{6}$.

8. $1\frac{5}{4}$, $\frac{7}{8}$, and $\frac{7}{18}$.   22. $\frac{2}{7}$, $\frac{3}{11}$, $\frac{4}{14}$, $\frac{7}{88}$, and $1\frac{2}{21}$.

9. $\frac{3}{11}$, $\frac{2}{14}$, and $1\frac{8}{6}$.   23. $\frac{8}{11}$, $\frac{17}{22}$, $\frac{114}{33}$, and $\frac{71}{6}$.

10. $\frac{11}{4}$, $\frac{8}{15}$, and $\frac{8}{66}$.   24. $\frac{112}{14}$, $\frac{19}{120}$, $\frac{81}{4}$, and $\frac{18}{66}$.

11. $\frac{10}{28}$, $\frac{6}{66}$, and $1\frac{30}{120}$.   25. $\frac{17}{24}$, $\frac{215}{84}$, $\frac{114}{66}$, and $\frac{11}{28}$.

12. $\frac{6}{60}$, $1\frac{8}{8}$, and $1\frac{9}{8}$.   26. $\frac{111}{21}$, $\frac{85}{33}$, $\frac{14}{4}$, and $1\frac{84}{4}$.

13. $\frac{9}{22}$, $\frac{4}{121}$, and $\frac{5}{143}$.   27. $\frac{84}{44}$, $\frac{84}{44}$, $\frac{10}{281}$, and $\frac{147}{13}$.

14. $\frac{3}{34}$, $\frac{121}{51}$, and $\frac{5}{68}$.   28. $4\frac{3}{8}$, $\frac{81}{33}$, $\frac{16}{4}$, and $\frac{5}{143}$.

15. $\frac{17}{16}$, $\frac{11}{128}$, and $\frac{9}{144}$.   29. $5\frac{3}{8}$, $11\frac{4}{21}$, $\frac{7}{11}$, and $\frac{90}{33}$.

16. $1\frac{1}{3}$, $1\frac{5}{22}$, and $2\frac{8}{48}$.   30. $6\frac{3}{7}$, $11\frac{8}{5}$, $4\frac{3}{13}$, and $\frac{84}{31}$.

NOTE 1.—If there are any integral, or any mixed numbers, reduce them to the form of simple fractions (Arts. 67 and 68).

NOTE 2.—Complex fractions are reduced to simple ones by the rule for division of fractions (Art. 81).

ADDITION OF FRACTIONS.
DEFINITION.

74. Addition of Fractions is the operation of finding the sum of two or more fractions.

MENTAL EXERCISES.

1. What is the fractional unit of $\frac{3}{11}$ of \$1 ? of $\frac{5}{11}$ of \$1 ? How many times is this unit taken in \$$\frac{3}{11}$ and \$$\frac{5}{11}$?

2. What is the sum of $\frac{5}{17}$ and $\frac{4}{17}$? of $\frac{5}{17}$, $\frac{3}{17}$, and $\frac{8}{17}$? of $\frac{2}{19}$, $\frac{5}{19}$, $\frac{4}{19}$ and $\frac{6}{19}$? of $\frac{7}{25}$, $\frac{4}{25}$, $\frac{11}{25}$, and $\frac{12}{25}$?

3. How many *twelfths* are there in $\frac{2}{3}$? How many *twelfths* in $\frac{3}{4}$? What is the sum $\frac{8}{12}$ and $\frac{9}{12}$? What is the sum of $\frac{2}{3}$ and $\frac{3}{4}$? What is the sum of $\frac{2}{3}$ and $\frac{4}{4}$?

NOTE.—The fractions must be reduced to the same fractional unit before they can be added, that is, they must be reduced to a *common denominator.*

4. A man bought $\frac{1}{2}$ of a *pound* of indigo at one time and $\frac{3}{8}$ of a *pound* at another time? how much did he buy in all? What is the sum of $\frac{1}{2}$ and $\frac{3}{8}$? of $\frac{4}{8}$ and $\frac{3}{8}$?

5. What is the sum of $\$4\frac{1}{4}$ and $\$2\frac{1}{4}$? of $1\frac{7}{4}$ and $\frac{7}{4}$? of $\frac{11}{12}$ and $\frac{28}{12}$? What mixed number is equal to $\frac{18}{12}$?

6. What is the sum of $\$\frac{2}{5}$, $\$\frac{1}{6}$, and $\$\frac{3}{10}$? of $\frac{2}{5}$, $\frac{1}{6}$, and $\frac{3}{10}$? of $\frac{22}{30}$, $\frac{6}{30}$, and $\frac{9}{30}$? What mixed number is equal to $\$\frac{11}{30}$?

OPERATION OF ADDITION.

75. Let it be required to find the sum of $\frac{4}{5}$ and $\frac{3}{7}$.

EXPLANATION.—Having reduced the fractions to a common denominator, we see that the first is equal to

OPERATION.

$$\frac{4}{5}+\frac{3}{7}=\frac{28}{35}+\frac{15}{35}=\frac{43}{35}=1\frac{8}{35}.$$

the fractional unit $\frac{1}{35}$ taken 28 times, and the second to the same unit taken 15 times; hence, the sum is equal to this unit taken 28 + 15, or 43 times, that is, to $\frac{43}{35}$ or to $1\frac{8}{35}$.

In like manner we may treat all similar cases; hence, the following

RULE.

I. Reduce the fractions to simple fractions having a common denominator.

II. Add their numerators for a new numerator, and write the sum over the common denominator.

MENTAL EXERCISES.

1. What is the sum of $\$\frac{7}{16}$, $\$\frac{19}{21}$, and $\$\frac{18}{35}$? *Ans.* $\$1\frac{7}{8}$.

Add the following groups of fractions:

2. $\frac{2}{3}$, $\frac{5}{6}$, and $\frac{7}{12}$.
3. $\frac{3}{5}$, $\frac{2}{7}$, and $\frac{1}{3}$.
4. $\frac{2}{5}$, $\frac{3}{8}$, and $\frac{7}{12}$.
5. $\frac{4}{7}$, $\frac{6}{25}$, and $\frac{9}{35}$.
6. $\frac{5}{12}$, $\frac{1}{8}$, $\frac{7}{24}$, and $\frac{1}{2}$.
7. $\frac{3}{4}$, $\frac{4}{5}$, and $\frac{7}{6}$.
8. $\frac{1}{2}$, $\frac{1}{3}$, $\frac{1}{4}$, and $\frac{1}{5}$.

9. $\frac{3}{8}$, $1\frac{5}{7}$, and $1\frac{33}{10}$.
10. $\frac{3}{5}$, $\frac{4}{15}$, and $1\frac{83}{20}$.
11. $\frac{4}{5}$, $\frac{7}{10}$, $\frac{5}{8}$, and $1\frac{1}{8}$.
12. $\frac{4}{7}$, $\frac{5}{14}$, $1\frac{1}{21}$, and $\frac{2}{3}$.
13. $\frac{5}{6}$, $\frac{8}{27}$, $\frac{1}{3}$, $\frac{4}{5}$, and $\frac{5}{6}$.
14. $\frac{7}{12}$, $\frac{1}{6}$, $\frac{3}{8}$, $1\frac{1}{4}$, and $\frac{1}{5}$.
15. $\frac{4}{7}$, $\frac{9}{14}$, $\frac{4}{5}$, $\frac{2}{21}$, and $\frac{1}{5}$.

When there are mixed numbers, add the sum of the fractional parts to the sum of the whole numbers.

16. $4\frac{1}{2}$, $6\frac{1}{4}$, $2\frac{1}{4}$, and $\frac{5}{6}$.　　　　*Ans.* $12 + \frac{23}{12} = 13\frac{11}{12}$.
17. $10\frac{3}{8}$, $7\frac{1}{4}$, $8\frac{2}{3}$, and $16\frac{3}{4}$.　　*Ans.* $42\frac{17}{36}$.
18. $1\frac{1}{10}$, $6\frac{2}{5}$, $18\frac{11}{8}$, and $2\frac{7}{30}$.　*Ans.* $28\frac{11}{60}$.
19. $2\frac{8}{11}$, $6\frac{1}{2}$, and $12\frac{12}{22}$.
20. $67\frac{3}{11}$, $4\frac{9}{33}$, and $600\frac{2}{3}$.
21. $13\frac{1}{7}$, $99\frac{3}{8}$, and $512\frac{5}{14}$.
22. $14\frac{6}{15}$, $3\frac{14}{15}$, and $88\frac{9}{15}$.
23. $900\frac{1}{10}$, $450\frac{37}{100}$, and $6\frac{9}{10}$.
24. $21\frac{1}{4}$, $98\frac{7}{16}$, and $14\frac{11}{24}$.
25. $1\frac{1}{8}$, $4\frac{1}{8}$, and $6\frac{2}{3}$.
26. $4\frac{4}{5}$, $7\frac{5}{12}$, and $8\frac{5}{6}$.
27. $\frac{68}{9}$, $\frac{4}{15}$, and $7\frac{7}{30}$.
28. $5\frac{1}{4}$, $\frac{3}{8}$, and $7\frac{11}{12}$.
29. $4\frac{5}{6}$, $\frac{73}{135}$, and $3\frac{1}{4}$.

30. $2\frac{3}{10}$, $5\frac{11}{15}$, and $6\frac{7}{20}$.
31. $5\frac{5}{11}$, $4\frac{3}{4}$, and $13\frac{1}{4}$.
32. $6\frac{4}{7}$, $11\frac{1}{14}$, and 9.
33. $\frac{11}{12}$, $\frac{5}{14}$, and $2\frac{5}{42}$.
34. $7\frac{1}{4}$, $11\frac{11}{12}$, and $5\frac{4}{5}$.
35. $18\frac{1}{4}$, $2\frac{5}{18}$, and $6\frac{1}{4}$.
36. $1\frac{5}{6}$, $1\frac{3}{34}$, and $8\frac{1}{51}$.
37. $9\frac{7}{11}$, $\frac{3}{4}$, and $\frac{17}{44}$.
38. $5\frac{1}{4}$, $3\frac{5}{18}$, and $\frac{11}{3}$.
39. $16\frac{1}{4}$, $2\frac{1}{8}$, and $\frac{81}{14}$.
40. $8\frac{8}{13}$, $\frac{23}{38}$, and $\frac{15}{16}$.

41. $\$\frac{3}{8} + \$\frac{4}{7} + \$\frac{9}{11} + \$\frac{1}{4} + \$3\frac{1}{2} = ?$
42. $37\frac{1}{2}$ *lbs.* $+ 24\frac{2}{4}$ *lbs.* $+ \frac{9}{13}$ *lb.* $+ 4\frac{1}{7}$ *lbs.* $= ?$
43. $46\frac{3}{4} + 118\frac{2}{7} + 319\frac{4}{5} + 1\frac{13}{16} = ?$
44. $39\frac{2}{5}$ *yds.* $+ 87\frac{1}{10}$ *yds.* $+ 17\frac{3}{20}$ *yds.* $+ 82\frac{11}{16}$ *yds.* $= ?$

PRACTICAL PROBLEMS.

1. A farmer has 3 fields; the first contains $31\frac{3}{8}$ *acres*, the second $49\frac{4}{7}$ *acres*, and the third $59\frac{7}{10}$ *acres*: how many acres has he in all? *Ans.* $140\frac{181}{280}$ *Acres.*

2. A man earned $\$3\frac{3}{8}$ the first day, $\$4\frac{1}{8}$ the second day, $\$5\frac{5}{16}$ the third day, $\$7\frac{1}{2}$ the fourth day, $\$4\frac{7}{16}$ the fifth day, and $\$3\frac{5}{8}$ the sixth day; how much did he earn in the six days? *Ans.* $\$28\frac{3}{8}$.

3. A. traveled $17\frac{2}{7}$ *miles* the first day, $\frac{3}{4}$ of $17\frac{2}{7}$ *miles* the second day, $22\frac{7}{8}$ *miles* the third day, and $36\frac{1}{4}$ *miles* the fourth day; how far did he travel in the four days?

4. B. works $8\frac{3}{5}$ *hours* on Monday, $9\frac{2}{3}$ *hours* on Tuesday, $8\frac{3}{10}$ *hours* on Wednesday, $10\frac{3}{4}$ *hours* on Thursday, $9\frac{4}{5}$ *hours* on Friday, and $10\frac{1}{3}$ *hours* on Saturday; how many hours does he work during the week?

5. A farmer sells 3 tons of hay for $\$47.18\frac{3}{4}$, 3 cows for $\$111.42\frac{1}{4}$, a horse for $\$173.16\frac{1}{6}$, and 100 bushels of oats for $\$62.87\frac{1}{2}$; how much does he receive for all?

6. How many dollars will pay for a coat worth $\$14\frac{3}{4}$, a hat worth $\$5\frac{1}{2}$, a vest worth $\$6\frac{1}{4}$, a pair of pants worth \$8, and a pair of boots worth $\$9\frac{1}{4}$?

7. How many *pounds* of butter in 4 tubs weighing respectively $27\frac{1}{3}$ *lbs.*, $34\frac{3}{8}$ *lbs.*, $32\frac{4}{5}$ *lbs.*, and $29\frac{5}{8}$ *lbs.*?

8. How many *tons* of coal in 5 loads weighing respectively $1\frac{1}{8}$, $1\frac{2}{11}$, $1\frac{3}{10}$, $\frac{12}{13}$, and $1\frac{1}{4}$ *tons*?

9. How many *yards* in 4 pieces of cloth, measuring respectively $27\frac{1}{2}$ *yds.*, $37\frac{3}{4}$ *yds.*, $39\frac{1}{5}$ *yds.*, and $30\frac{11}{13}$ *yds.*?

10. A farm contains $26\frac{1}{2}$ *acres* of plough land, $39\frac{1}{4}$ *acres* of wood land, $61\frac{3}{8}$ *acres* of pasture land, and $42\frac{1}{8}$ *acres* of meadow land; how many *acres* in the farm?

SUBTRACTION OF FRACTIONS.

DEFINITIONS.

76. **Subtraction of Fractions** is the operation of finding the **difference** between two fractions.

MENTAL EXERCISES.

1. What is the difference between $\frac{6}{8}$ and $\frac{3}{8}$? $\frac{11}{13}$ and $\frac{7}{13}$? $\frac{14}{17}$ and $\frac{7}{17}$? $\frac{22}{33}$ and $\frac{17}{33}$? $\frac{14}{24}$ and $\frac{12}{24}$? $\frac{14}{19}$ and $\frac{8}{19}$?

2. .What is the difference between $\frac{5}{6}$ and $\frac{2}{5}$?

EXPLANATION.—The minuend is equal to $\frac{25}{30}$ and the subtrahend is equal to $\frac{12}{30}$; the difference is therefore equal to the fractional unit $\frac{1}{30}$ taken, $25-12$, or 13 times, that is, it is to $\frac{13}{30}$.

3. A boy had $\frac{7}{8}$ of a *dollar*, but he spent $\frac{3}{16}$ of a dollar for a slate; how much had he left? What is the difference between $\frac{7}{8}$ and $\frac{3}{16}$? $\frac{14}{16}$ and $\frac{3}{16}$?

4. $\frac{4}{5} - \frac{3}{8} = ?$ 6. $\frac{17}{4} lbs. - \frac{3}{8} lb. = ?$

5. $\$1\frac{4}{5} - \$\frac{3}{11} = ?$ 7. $2\frac{9}{10} ft. - \frac{11}{12} ft. = ?$

OPERATION OF SUBTRACTION.

77. Let it be required to find the difference between $\frac{5}{8}$ and $\frac{2}{7}$.

EXPLANATION.—Having reduced the given fractions to the common unit, $\frac{1}{56}$, we see that the minuend contains this unit 35 times and the subtrahend 16 times; hence, the remainder contains it $35 - 16$, or 19 times; the remainder is therefore $\frac{19}{56}$.

OPERATION.

$$\frac{5}{8} - \frac{2}{7} = \frac{35}{56} - \frac{16}{56} = \frac{19}{56}.$$

Since all similar examples may be treated in the same manner, we have the following

RULE.

I. Reduce the fractions to simple fractions having a common denominator.

II. Subtract the numerator of the subtrahend from that of the minuend for a new numerator, and write the difference over the common denominator.

EXAMPLES.

Find the difference between,

1. $\frac{11}{18}$ and $\frac{11}{12}$.
2. $\frac{8}{15}$ and $\frac{9}{20}$.
3. $37\frac{4}{15}$ and $33\frac{5}{24}$.
4. $6\frac{3}{4}$ and $4\frac{1}{2}$.
5. $13\frac{5}{12}$ and $9\frac{4}{12}$.
6. $50\frac{1}{16}$ and $47\frac{1}{24}$.
7. 42 and $30\frac{5}{12}$.
8. $90\frac{8}{11}$ and $25\frac{9}{22}$.
9. $46\frac{6}{8}$ and $15\frac{1}{4}$.
10. $\frac{3}{10}$ and $\frac{2}{3}$ of $\frac{3}{4}$.
11. $98\frac{1}{2}$ and $45\frac{3}{4}$.
12. $150\frac{1}{20}$ and $65\frac{1}{10}$.
13. $8\frac{1}{2}$ and $\frac{4}{5}$.
14. $3\frac{7}{15}$ and $1\frac{3}{8}$.
15. $13\frac{1}{4}$ and $7\frac{11}{12}$.
16. $843\frac{24}{31}$ and $94\frac{1}{4}$.

17. \$$18\frac{15}{16}$ and \$$13\frac{11}{12}$.
18. $14\frac{3}{11}ft.$ and $7\frac{4}{7}ft.$
19. $10\frac{81}{200}$ and $4\frac{11}{16}$.
20. $206\frac{1}{4}yds.$ and $194\frac{4}{5}yds.$
21. $47\frac{23}{24}yds.$ and $27\frac{11}{14}yds.$
22. \$$118\frac{15}{16}$ and \$$24\frac{11}{12}$.
23. \$$22\frac{14}{100}$ and \$$9\frac{77}{100}$.
24. $246\frac{11}{16}$ and $194\frac{4}{7}$.
25. $1,476\frac{23}{24}$ and $894\frac{3}{11}$.
26. $177\frac{3}{4}bu.$ and $66\frac{11}{12}bu.$
27. $163\frac{3}{8}mi.$ and $108\frac{5}{8}mi.$
28. \$$864\frac{7}{16}$ and \$$648\frac{3}{4}$.
29. $146\frac{3}{4}tons$ and $97\frac{7}{8}tons.$
30. $221\frac{1}{8}acres$ and $141\frac{3}{4}acres.$
31. $1,884\frac{3}{11}in.$ and $1,801\frac{4}{5}in.$
32. $280\frac{1}{10}rods$ and $199\frac{2}{15}rods.$

33. \$$180\frac{1}{4} + \$27\frac{1}{2} - (\$43\frac{3}{8} + \$39\frac{3}{4}) = ?$
34. $146\frac{1}{2}rods + 73\frac{1}{4}rods - (24\frac{1}{3}rods - 6\frac{1}{4}rods) = ?$
35. $118bu. + 375\frac{3}{4}bu. - (46bu. - 7\frac{1}{2}bu.) = ?$

PRACTICAL PROBLEMS.

1. A. has \$$4\frac{7}{16}$, and B. has \$$3\frac{15}{16}$; how much more has A. than B.?   *Ans.* \$$\frac{1}{2}$.

2. A. bought $56\frac{3}{10}$ *pounds* of butter, and sold $\frac{1}{2}$ of it to

4

one person and $13\frac{4}{7}$ *pounds* to another; how much had he left? *Ans.* $14\frac{121}{140}$ *pounds.*

3. A grocer bought 2 *hogsheads* of sugar, each weighing 1,302 *pounds;* he sold $\frac{1}{3}$ of one hogshead and $455\frac{2}{11}$ *pounds* from the other: how many pounds had he remaining? *Ans.* $1{,}714\frac{9}{11}$ *lbs.*

4. A merchant bought 2 pieces of cloth; the first contained $38\frac{3}{8}$ *yards,* and the second $41\frac{4}{7}$ *yards;* he then sold $59\frac{1}{4}$ *yards:* how many yards had he left?

5. A man bought a farm of $211\frac{3}{4}$ *acres,* and sold $117\frac{5}{11}$ *acres* of it; how much had he remaining?

6. From a cask containing $60\frac{2}{3}$ *gallons* of cider there were drawn off $17\frac{11}{13}$ *gallons;* how much was there left?

7. A sloop has on board $406\frac{3}{8}$ *tons* of coal, of which $311\frac{4}{9}$ *tons* is anthracite, and the remainder cannel; how much cannel does she contain?

8. A merchant had a piece of silk containing $42\frac{1}{4}$ *yards,* from which he sold $17\frac{3}{4}$ *yards;* how much had he remaining?

9. A person having \$$49\frac{3}{4}$, spent \$$4\frac{3}{4}$ in getting to Boston, and \$$5\frac{7}{10}$ in getting to Portland; how much had he left on reaching Portland?

10. If a man has \$$11\frac{1}{2}$ and spends \$$8\frac{3}{4}$, how much will he have left?

11. A tailor had a piece of cloth containing $29\frac{1}{2}$ *yds;* he cut off $4\frac{1}{4}$ *yds.* to make a coat and $1\frac{3}{4}$ *yds.* to make a pair of pants: how many yards were left?

12. A man had to walk $97\frac{3}{4}$ *miles;* he walked $30\frac{3}{4}$ *miles* the first day, $33\frac{1}{2}$ *miles* the second day, and finished the journey the third day: how far did he walk the third day?

13. A cask of wine contained $42\frac{1}{2}$ *gallons*; of this $13\frac{3}{4}$ *gallons* were drawn off, and $12\frac{1}{4}$ *gallons* leaked out: how much remained in the cask?

14. A grocer bought $89\frac{1}{2}$ *lbs.* of tea; of this he sold $13\frac{1}{4}$ *lbs.* to one customer, $9\frac{1}{2}$ *lbs.* to a second customer, and $12\frac{3}{4}$ *lbs.* to a third customer: how many pounds had he left?

15. A laborer earned $\$18\frac{3}{8}$ and received as a gift $\$14\frac{1}{4}$; he then bought a barrel of flour for \$12, and groceries to the amount of $\$8\frac{11}{12}$: how much had he remaining?

16. A drover bought 4 cows for $\$168\frac{1}{4}$, and after paying $\$34\frac{1}{10}$ for pasturage, he sold them for $\$203\frac{1}{2}$; did he gain or lose, and how much?

17. A man traveled in a certain direction $34\frac{2}{11}$ *miles* the first day, and $37\frac{3}{8}$ *miles* the second day; then, retracing his path, he traveled $28\frac{3}{11}$ *miles* on the third day, and $34\frac{1}{8}$ *miles* on the fourth day: how far was he then from the starting point?

18. What is the difference between $57\frac{1}{4}$ *yds.* + $72\frac{3}{13}$ *yds.* and 211 *yds.* — $94\frac{2}{15}$ *yds.*?

19. A. bought a house for $\$11,320\frac{1}{2}$, and after expending $\$1,311\frac{3}{5}$ for repairs, sold it for \$12,500; did he gain or lose, and how much?

MULTIPLICATION OF FRACTIONS.

DEFINITION.

78. Multiplication of Fractions is the operation of finding the **product** of two or more fractions.

MENTAL EXERCISES.

1. What is $\frac{1}{5}$ of 5? $\frac{2}{5}$ of 5? $\frac{3}{5}$ of 5? What is 5 taken $\frac{1}{5}$ of 1 time? 5 taken $\frac{2}{5}$ of one time? 5 taken $\frac{3}{5}$ of 1 time? What is $5 \times \frac{1}{5}$? $5 \times \frac{2}{5}$? $5 \times \frac{3}{5}$? $5 \times \frac{4}{5}$? $5 \times \frac{13}{15}$?

2. If $\frac{1}{3}$ of an orange is divided into 5 equal parts, how much of a whole orange is 1 of these parts? how much is 2 of the parts? 3 of the parts? What is $\frac{1}{5}$ of $\frac{1}{3}$? $\frac{2}{5}$ of $\frac{1}{3}$? $\frac{3}{5}$ of $\frac{1}{3}$? What is $\frac{1}{3} \times \frac{1}{5}$? $\frac{1}{3} \times \frac{2}{5}$? $\frac{1}{3} \times \frac{3}{5}$? $\frac{1}{3} \times \frac{4}{5}$?

EXPLANATION.—The expression $\frac{1}{5}$ of $\frac{1}{3}$ is equivalent to the expression $\frac{1}{3} \times \frac{1}{5}$; for $\frac{1}{5}$ of $\frac{1}{3}$ is the same as $\frac{1}{3}$ taken $\frac{1}{5}$ of 1 time; it is therefore equal to $\frac{1}{3} \times \frac{1}{5}$ (Art. **27**).

3. What is $\frac{1}{4}$ of $\frac{1}{5}$? $\frac{2}{4}$ of $\frac{1}{5}$? $\frac{3}{4}$ of $\frac{1}{5}$? What is $\frac{1}{4}$ of $\frac{3}{5}$? $\frac{2}{4}$ of $\frac{3}{5}$? What is $\frac{1}{4} \times \frac{1}{5}$? $\frac{2}{4} \times \frac{1}{5}$? $\frac{1}{4} \times \frac{3}{5}$? $\frac{2}{4} \times \frac{3}{5}$? Is there any difference between $\frac{2}{4}$ of $\frac{3}{7}$, and $\frac{3}{7}$ of $\frac{2}{4}$? What is $\frac{3}{8} \times \frac{5}{6}$? What is $\frac{4}{5} \times \frac{5}{7}$? $\frac{3}{11} \times \frac{5}{7}$?

OPERATION OF MULTIPLICATION.

79. Let it be required to multiply $\frac{3}{4}$ by $\frac{5}{7}$.

EXPLANATION.—We first multiply $\frac{3}{4}$ by 5, which, according to Principle 1° (Art. **65**) gives $\frac{3 \times 5}{4}$; but

OPERATION.

$$\frac{3}{4} \times \frac{5}{7} = \frac{3 \times 5}{4 \times 7} = \frac{15}{28}$$

this result is 7 times the required product, because the multiplier used is 7 times the given multiplier; hence, to find the true product, we must divide it by 7, which, according to Principle 3° (Art. **65**), gives $\frac{3 \times 5}{4 \times 7}$, or $\frac{15}{28}$.

In like manner we may treat all similar cases; hence, the following

RULE.

Reduce the factors to the form of simple fractions; then multiply the numerators together for a new numerator, and the denominators for a new denominator.

EXAMPLES.

1. Multiply $7\frac{1}{2}$ by $\frac{3}{5}$.   *Ans.* $\dfrac{\overset{3}{\cancel{15}}}{2} \times \dfrac{3}{5} = \dfrac{9}{2} = 4\frac{1}{2}$.

2. Multiply $2\frac{1}{4}$ by $\frac{1}{5}$ of $\frac{4}{9}$.   *Ans.* $\dfrac{9}{4} \times \dfrac{1}{5} \times \dfrac{4}{9} = \dfrac{1}{5}$.

NOTE.—After indicating the operation, *cancel every factor that is* common to any numerator and any denominator. If the final result is an improper fraction, reduce it to a whole, or to a mixed number; if it is a proper fraction, reduce it to its lowest terms.

The rule may be simplified in the following cases:

1°. To multiply a whole number by a simple fraction:

Multiply it by the numerator of the fraction and divide the result by the denominator.

3. Multiply 928 by $\frac{3}{8}$.   *Ans.* $\frac{928}{1} \times \frac{3}{8} = \frac{928 \times 3}{8} = 348.$

2°. To multiply a whole number by a mixed number:

Multiply it first by the fractional part of the mixed number, then by the integral part, and find the sum of the results.

(4.)	928		(5.)	3)1143	
	$6\frac{3}{8}$			$7\frac{1}{3}$	
	348	$\frac{3}{8}$ of 928.		381	$\frac{1}{3}$ of 1143.
	5568	6 times 928.		8001	7 times 1143.
	5916	Product.		8382	Product.

Perform the following indicated operations:

6. $\frac{8}{5} \times \frac{21}{160}.$

7. $\frac{80}{9} \times \frac{9}{4}.$

8. $\frac{7}{8} \times 12\frac{4}{5}.$

9. $7\frac{1}{2} \times 8\frac{3}{4}.$

10. $7\frac{1}{3} \times 61\frac{3}{5}.$

11. $\frac{5}{8}$ of $7 \times \frac{3}{5}.$

12. $15\frac{4}{5} \times \frac{21}{5}.$

13. $6\frac{2}{3} \times \frac{162}{16}.$

14. $\frac{6}{11}$ of $\frac{2}{3} \times \frac{12}{24}.$

15. $\frac{3}{4}$ of $7\frac{1}{5} \times \frac{1}{4}$ of 90.

16. $345 \times 4\frac{7}{9}.$

17. $3\frac{7}{9} \times 2\frac{2}{8}.$

18. $\frac{8}{5} \times \frac{3}{8} \times \frac{7}{11}.$

19. $3\frac{1}{2} \times \frac{7}{8} \times 11\frac{1}{2}.$

20. $\frac{7}{18} \times \frac{3}{14} \times \frac{36}{5}.$

21. $3\frac{1}{8} \times \frac{20}{187} \times \frac{34}{5}.$

22. $3\frac{1}{3} \times 7\frac{1}{2} \times \frac{27}{23}.$

23. $\frac{64}{11} \times 2\frac{1}{4} \times 1\frac{4}{15}.$

24. $5\frac{1}{4} \times 10\frac{1}{8} \times 3\frac{1}{4}$.

25. $2\frac{2}{5} \times 5\frac{5}{8} \times \frac{122}{315}$.

26. $114\frac{113}{215} \times 81\frac{1}{2}$.

27. $\frac{3}{4}$ of $\frac{5}{11} \times 1\frac{7}{20}$.

28. $2\frac{1}{8} \times \frac{50}{51} \times \frac{3}{5}$.

29. $\frac{7}{8}$ of $3\frac{1}{3} \times \frac{14}{17}$.

30. $1\frac{1}{13} \times \frac{7}{8} \times 3\frac{1}{5}$.

31. $217\frac{1}{4} \times 112\frac{5}{8}$.

32. $\frac{14}{21} \times \frac{7}{8} \times \frac{3}{17}$.

33. $8\frac{1}{4} \times 8\frac{1}{4} \times 8\frac{1}{4}$.

34. $(56\frac{7}{8} + 24\frac{1}{2}) \times (13\frac{1}{4} + 9\frac{3}{8}) = ?$

35. $(111 + 302\frac{1}{4}) \times (107\frac{1}{4} - 30\frac{3}{14}) = ?$

36. $(207\frac{2}{3} + 39\frac{1}{6}) \times (100\frac{1}{11} - 66\frac{3}{11}) = ?$

37. $(445\frac{2}{8} - 36\frac{1}{4}) \times (36\frac{3}{4} - 21\frac{7}{12}) = ?$

38. $(999\frac{2}{7} - \frac{246}{7}) \times (\frac{384}{3} - 72\frac{1}{8}) = ?$

39. $(256 - 7\frac{1}{3}) \times (394 + 4\frac{2}{8}) = ?$

40. $(224 + 3\frac{4}{11}) \times (88\frac{2}{8} - 4\frac{1}{10}) = ?$

PRACTICAL PROBLEMS.

1. If a man earns \$33⅓ per week, how much will he earn in a year of 52 weeks ?   *Ans.* \$1,733⅓.

2. A farmer bought 43 acres of land at \$104¼ per acre, 16 cows at \$28⅘ each, and 2 plows at \$11$\frac{6}{13}$ each ; what did they all cost him ?   *Ans.* \$4,985$\frac{9}{65}$.

3. What must be paid for 600 barrels of flour at \$5.37½ per barrel ?   *Ans.* \$3,225.

4. What is the cost of 33⅓ lbs. of tea at 93¾ cents a pound ?   *Ans.* \$31.25.

5. If a man can travel 7¾ miles in 1 hour, how many miles can he travel in 6⅓ hours ?

6. If it takes 1¼ bushels of wheat to sow an acre, how many bushels will it take to sow 7$\frac{9}{10}$ acres ?

7. A grocer bought 100 barrels of flour at \$6¾ per barrel ; he sold 49 barrels at \$7½ per barrel, and the rest at \$7⅓ per barrel : how much did he gain ?

8. A. bought $319\frac{3}{7}$ acres of land at \$200 per acre; he then sold $250\frac{4}{5}$ acres at \$250 per acre, and the remainder at $\$266\frac{2}{3}$ per acre: how much did he gain?

9. A drover bought 64 sheep at $\$7\frac{3}{4}$ a piece; he then sold 30 of them at $\$6\frac{7}{8}$ a piece, and the remainder at $\$8\frac{7}{8}$ a piece: did he gain or lose, and how much?

10. A. starts from Cincinnati and travels at the rate of $5\frac{3}{4}$ miles an hour; at the end of $3\frac{1}{2}$ hours B. starts from the same place and travels in pursuit at the rate of $6\frac{1}{4}$ miles an hour: how far apart are they at the end of $5\frac{1}{4}$ hours?

11. What is the cost of $\frac{3}{7}$ of a piece of cloth containing $13\frac{1}{2}$ yards at $\$2\frac{1}{2}$ a yard?

12. A., B., and C. own a tract of land; A's share is $62\frac{1}{2}$ acres, B's share is $1\frac{1}{2}$ times as much as A's, and C's share is $10\frac{1}{4}$ acres greater than A's and B's together: how many acres in the whole tract?

13. A man traveled $112\frac{1}{2}$ miles in 3 days; the first day he traveled $\frac{2}{5}$ of the whole distance, and the second day he traveled $\frac{7}{8}$ of the distance he did the first day: how far did he travel the third day?

14. A woman is $24\frac{3}{4}$ years old, and her husband lacks $7\frac{3}{4}$ years of being twice as old; what are the united ages of the two?

15. What will $\frac{3}{8}$ of $\frac{5}{8}$ of a yard of cloth cost at the rate of $\frac{8}{17}$ of $\$3\frac{2}{5}$ per yard?

16. How many yards in 8 pieces of cloth, each containing $37\frac{2}{3}$ yards?

17. If a train of cars runs $22\frac{1}{3}$ *mi.* an hour, how far will it run in $8\frac{1}{4}$ *hrs?*

DIVISION OF FRACTIONS.

DEFINITION.

80. **Division of Fractions** is the operation of find-ing the **quotient** of one fraction by another.

MENTAL EXERCISES.

1. In 1 *orange,* how many *fifths of an orange?* How many times is $\frac{1}{5}$ contained in 1 ? What is the quotient of 1 by $\frac{1}{5}$? of 1 by $\frac{1}{6}$? of 1 by $\frac{1}{7}$?

2. What is the quotient of 1 by $\frac{1}{7}$? of 2 by $\frac{1}{7}$? of 3 by $\frac{1}{7}$? How do you divide an integral number by a fractional unit ?

EXPLANATION.—The quotient of 1 by $\frac{1}{7}$ is 7 ; but the quotient of 3 by $\frac{1}{7}$ is 3 times as great as the quotient of 1 by $\frac{1}{7}$; hence it is 3×7, that is, *we multiply the given number by the denominator of the frac-tional unit.*

3. What is the quotient of 3 by $\frac{1}{7}$? of $\frac{3}{4}$ by $\frac{1}{7}$? of $\frac{3}{4}$ by $\frac{1}{7}$? How do you divide a simple fraction by a fractional unit ?

EXPLANATION.—The quotient of 3 by $\frac{1}{7}$ is 3×7 ; but the quotient of $\frac{3}{4}$ by $\frac{1}{7}$ is only $\frac{1}{4}$ as great as the quotient of 3 by $\frac{1}{7}$; hence it is $\frac{3 \times 7}{4}$, that is, *we multiply the numerator of the given fraction by the denominator of the fractional unit.*

4. What is the quotient of $\frac{3}{4}$ by 1 ? of $\frac{3}{4}$ by 2 ? of $\frac{3}{4}$ by 5 ? How do you divide a simple fraction by a whole num-ber ?

EXPLANATION.—The quotient of $\frac{3}{4}$ by 1 is $\frac{3}{4}$; but the quotient of $\frac{3}{4}$ by 5 is only $\frac{1}{5}$ as great as the quotient of $\frac{3}{4}$ by 1 ; hence it is $\frac{3}{4} \times \frac{1}{5}$ or $\frac{3}{4 \times 5}$, that is, *we multiply the denominator of the given fraction by the whole number.*

OPERATION OF DIVISION.

81. Let it be required to divide $\frac{3}{7}$ by $\frac{4}{5}$:

EXPLANATION.—Here the divisor is equal to $\frac{1}{5}$ taken 4 times, that is, to $\frac{1}{5} \times 4$; hence, to find the quotient we divide $\frac{3}{7}$ by $\frac{1}{5}$ and that result by 4. To divide $\frac{3}{7}$ by $\frac{1}{5}$ we multiply its numerator by 5 (Art. **80**, Ex. 3), which gives $\frac{3 \times 5}{7}$;

OPERATION.

$$\frac{3}{7} \div \frac{4}{5} = \frac{3}{7} \div \left(\frac{1}{5} \times 4\right)$$

$$= \frac{3 \times 5}{7 \times 4} = \frac{3}{7} \times \frac{5}{4} = \frac{15}{28}.$$

to divide the result by 4 we multiply its denominator by 4, (Art. **80**, Ex. 3), which gives $\frac{3 \times 5}{7 \times 4}$, and this is the same thing as $\frac{3}{7} \times \frac{5}{4}$, or $\frac{15}{28}$.
Here we have inverted the divisor, that is, we have made its terms change places, and then we have proceeded as in multiplication.

In like manner we may treat all similar cases ; hence, the

RULE.

Reduce both dividend and divisor to simple fractions; then invert the divisor and proceed as in multiplication.

NOTE.—Before performing the multiplication *cancel and reduce* as explained in Art. **79**.

EXAMPLES.

1. Divide $3\frac{1}{5}$ by $\frac{3}{5}$.   *Ans.* $\frac{16}{5} \times \frac{5}{3} = \frac{16}{3} = 5\frac{1}{3}$.

To divide a whole number by a simple fraction we may

Multiply it by the denominator of the fraction and divide the result by the numerator.

2. Divide 27 by $\frac{4}{5}$.   *Ans.* $\frac{27}{1} \times \frac{5}{4} = \frac{27 \times 5}{4} = 33\frac{3}{4}$.

Perform the following indicated operations :

3. $\frac{51}{64} \div \frac{17}{16}$.   4. $\frac{2}{3} \times \frac{7}{16} \div 86\frac{1}{16}$.

5. $\frac{3}{8} \div 1\frac{3}{16}$.

6. $7\frac{3}{16} \div 1\frac{9}{17}$.

7. $1\frac{23}{51} \div \frac{21}{34}$.

8. $241 \div \frac{27}{7}$.

9. $1275 \div \frac{25}{33}$.

10. $\frac{3}{4} \div \frac{3}{8} \times \frac{9}{11}$.

11. $\frac{19}{8} \div \frac{1}{7}$ of $\frac{3}{11}$.

12. $\frac{4}{5}$ of $1\frac{4}{4} \div 1\frac{8}{9} \times \frac{7}{15}$.

13. $\frac{2}{5}$ of $1\frac{9}{11} \div 1\frac{9}{11}$ of $\frac{1}{3}$.

14. $1\frac{14}{29} \div \frac{2}{3}$ of $\frac{3}{5}$.

15. $4\frac{11}{13} \div 1\frac{8}{7}$.

16. $\frac{563}{5} \div 21$.

17. $4\frac{31}{50} \div 4\frac{21}{7}$.

18. $54 \div 3\frac{3}{5}$.

19. $611\frac{7}{11} \div 20\frac{7}{11}$.

20. $100\frac{27}{30} \div 66\frac{2}{3}$.

21. $\frac{5}{6}$ of $1\frac{3}{28} \div \frac{3}{15}$ of $1\frac{9}{11}$.

22. $\frac{3}{4}$ of $7\frac{3}{4} \div \frac{1}{8}$ of $2\frac{1}{16}$.

23. $(8\frac{1}{4} + 3\frac{1}{2}) \div 7\frac{3}{8}$.

24. $(7\frac{3}{4} + 8\frac{1}{4}) \div \frac{1}{2}$ of $6\frac{2}{11}$.

25. $\frac{1}{8}$ of $4\frac{1}{2} \div (2\frac{1}{2} + 3\frac{2}{8})$.

26. $7\frac{3}{4} \times 8\frac{4}{7} \div 3\frac{1}{8} \times 2\frac{1}{2}$.

27. $1\frac{3}{8}$ of $3\frac{1}{2} \div 21\frac{1}{4} \times 7\frac{1}{4}$.

28. $(3\frac{1}{4} + 15\frac{3}{4}) \div 27\frac{3}{28}$.

29. $25\frac{4}{15} \div (7\frac{1}{3} + 5\frac{1}{4})$.

30. $11\frac{7}{11} \div (\frac{3}{22} + 7\frac{7}{11})$.

31. $14\frac{9}{10} \div \frac{7}{8}$ of 15.

32. $214\frac{3}{4} \div \frac{1}{8}$ of $25\frac{11}{13}$.

33. $\frac{1}{2}$ of $\frac{2}{3}$ of $5 \div 21\frac{11}{13}$.

34. $4 \times 7\frac{1}{2} \div 8 \times 19\frac{1}{4}$.

35. $\frac{1}{2}$ of $15\frac{3}{4} \div 9\frac{1}{2} \times \frac{4}{5}$.

36. $(32\frac{1}{4} + 7\frac{1}{4}) \div \frac{3}{7}$ of $\frac{7}{11}$.

37. $2\frac{7}{8} \times 3\frac{2}{7} + 3 \div 4\frac{1}{2} - 7\frac{1}{4} = ?$

38. $(3\frac{1}{12} + 7\frac{5}{8}) \div 2\frac{5}{16} + 7\frac{1}{2} \div 5\frac{2}{7} = ?$

39. $(11\frac{1}{4} + 17\frac{3}{4}) \div (33\frac{1}{8} - 4\frac{1}{8}) + 8\frac{7}{16} = ?$

40. $(\frac{1}{2}$ of $6\frac{3}{4} - 2\frac{1}{8}) \div (25 + \frac{1}{4}$ of $3\frac{1}{8}) = ?$

NOTE.—The following problems afford exercises on all the operations that can be performed upon fractions:

PRACTICAL PROBLEMS.

1. If $7\frac{1}{2}$ yds. of silk cost $13, what is the cost of 1 yard? of 5 yds.? *Ans.* $1\frac{11}{15}$; and $8\frac{2}{3}$.

2. If $37\frac{1}{2}$ ounces of silver cost $31\frac{1}{4}$, what is the cost of 1 ounce? of 150 ounces? *Ans.* $\frac{5}{6}$; and $125.

3. If $3\frac{3}{4}$ bu. of buckwheat cost $2\frac{3}{8}$, what does 1 bu. cost? What is the cost of 30 bu.? *Ans.* $\frac{19}{30}$; and $19.

4. A. divides $3,000\frac{3}{4}$ into 7 equal shares and gives $4\frac{1}{2}$

of these shares to a benevolent society; how much does he give to the society? *Ans.* $1,928$\frac{23}{33}$.

5. A. can build a wall in 10 days, B. can do it in 12 days, and C. in 15 days; what part of the wall can they all build in 1 day? *Ans.* $\frac{1}{10} + \frac{1}{12} + \frac{1}{15} = \frac{1}{4}$.

6. How long will it take them all to build the wall?

EXPLANATION.—Because it takes 1 day to build $\frac{1}{4}$ of the wall, it will take 4 days to build the whole.

7. What number multiplied by $1\frac{3}{8}$ will give $14\frac{3}{4}$?

8. The difference of two numbers is $15\frac{4}{35}$ and the greater number is $20\frac{11}{15}$; what is the less number?

9. A man inherits $\frac{2}{3}$ of an estate and gives his son $\frac{1}{2}$ of his share; what part of the estate does the son receive?

10. If $\frac{2}{3}$ of a ton of coal costs $13, what will 7 tons cost?

11. If coffee costs $13\frac{3}{4}$ cents a pound, how much can be bought for $10? for $16?

12. How many pounds of coffee can be bought for 78\frac{4}{5}$ at $\frac{3}{16}$ per pound? at $\frac{2}{3}$?

13. A. can do a piece of work in 3 days and B. can do it in 2 days; how long will it take them both to do it?

14. A. bought $24\frac{1}{2}$ *yds.* of cloth at $4\frac{1}{4}$ a yard, and sold the whole for $128\frac{3}{4}$; what did he gain per yard?

15. How many pounds of sugar at $12\frac{1}{2}$ *cts.* a pound must be given for $16\frac{1}{2}$ *lbs.* of butter at $22\frac{1}{2}$ *cts.* a pound?

16. If 6 men can do a piece of work in $7\frac{1}{2}$ days, how long will it take one man to do it?

17. If a man can walk $10\frac{4}{5}$ miles in $1\frac{1}{4}$ hours, how far can he walk in 1 hour? in $5\frac{4}{5}$ hours?

18. A merchant owning $\frac{4}{15}$ of a vessel, sold $\frac{1}{2}$ of his share for $1,640; what was the vessel worth at that rate?

19. A. can mow a piece of grass in 4 days, and B. can do it in 2 days; how long will it take both to do it?

20. A man made a journey in $6\frac{1}{2}$ days, traveling at the rate of $22\frac{3}{4}$ miles a day; on his return he traveled at the rate of $24\frac{1}{2}$ miles a day: how many days did it take him to return?

21. A merchant bought a piece of cloth containing $36\frac{1}{4}$ *yds.* for $\$65\frac{1}{8}$; at what rate must he sell it per yard so as to gain $\$25\frac{1}{2}$?

22. A. sets out from Detroit and travels towards Buffalo at the rate of $6\frac{3}{4}$ miles an hour; at the end of $2\frac{1}{2}$ hours B. sets out from Detroit and follows at the rate of $8\frac{1}{4}$ miles an hour: how far apart are they at the end of $5\frac{3}{4}$ hours?

23. A farmer sold to a grocer $32\frac{1}{2}$ *bu.* of corn at $\$\frac{7}{8}$ a bushel, and 86 *lbs.* of butter at $\$\frac{3}{8}$ a pound. He received in pay 200 *lbs.* of sugar at $\$\frac{1}{8}$ a pound, and the remainder in money; how much money did he receive?

24. A regiment lost 220 men in battle, which was 4 men more than $\frac{2}{3}$ of the whole regiment; how many men were there in the regiment?

25. A. owned $\frac{7}{8}$ of a ship and sold $\frac{3}{5}$ of his share to B.; B. then sold $\frac{2}{7}$ of what he bought to C. for $3000; what was the whole ship worth at that rate?

CONTRACTIONS IN MULTIPLICATION AND DIVISION.

82. The rules for multiplication and division of fractions lead to certain contractions in multiplication and division of whole numbers, of which the following are some of the most important.

1°. The fraction $\frac{100}{4}$ is equal to 25; hence, *to multiply a number by 25,* we may

Annex 2 ciphers and divide the result by 4.

To divide a number by 25, we may

Multiply it by 4 and divide the result by 100.

EXAMPLES.

1. Multiply 3,416 by 25. *Ans.* $\dfrac{341600}{4} = 85400.$

2. Divide 5,875 by 25. *Ans.* $\dfrac{5875 \times 4}{100} = 235.$

3. $394 \times 25 = ?$	8. $9,850 \div 25 = ?$
4. $3,724 \times 25 = ?$	9. $93,100 \div 25 = ?$
5. $8,123 \times 25 = ?$	10. $87,525 \div 25 = ?$
6. $10,201 \times 25 = ?$	11. $46,350 \div 25 = ?$
7. $4,386 \times 25 = ?$	12. $174,025 \div 25 = ?$

Let the pupil deduce rules for multiplying and dividing by $12\frac{1}{2}$, by $33\frac{1}{3}$, and by 125:

13. $81 \times 12\frac{1}{2} = ?$	21. $10,125 \div 12\frac{1}{2} = ?$
14. $914 \times 12\frac{1}{2} = ?$	22. $\$11,425 \div \$12\frac{1}{2} = ?$
15. $\$4,834 \times 12\frac{1}{2} = ?$	23. $9,125 \, yds. \div 12\frac{1}{2} = ?$
16. $\$375 \times 33\frac{1}{3} = ?$	24. $13,500 \div 33\frac{1}{3} = ?$
17. $28,452 \, ft. \times 33\frac{1}{3} = ?$	25. $\$29,100 \div 33\frac{1}{3} = ?$
18. $1,876 \, yds. \times 125 = ?$	26. $8,700 \, yds. \div 33\frac{1}{3} \, yds. = ?$
19. $\$4,365 \times 125 = ?$	27. $\$2,250 \div 125 = ?$
20. $34,115 \, yds. \times 12\frac{1}{2} = ?$	28. $10,000 \, ft. \div 125 = ?$

REVIEW QUESTIONS.

(60.) What is a fractional unit? What is a half, a third, a fourth, &c. ? What is the reciprocal of a number? **(61.)** What is a fraction? How do you write a common fraction? What is the denominator? the numerator? What are terms? **(63.)** In how many ways may we regard a fraction? Illustrate. **(64.)** What is a proper fraction? Illustrate. An improper fraction? Illustrate.

A mixed number? Illustrate. A simple fraction? Illustrate. A complex fraction? Illustrate. (**65.**) State the fundamental principles of fractions. (**66.**) What is reduction? (**67.**) Give the rule for reducing a whole number to a fraction with a given unit. (**68.**) Give the rule for reducing a mixed number to a fraction. (**69.**) Give the rule for reducing an improper fraction to a mixed number. (**70.**) Give the rule for reducing a fraction to its lowest terms. (**72.**) Give the rule for reducing fractions to a common denominator. (**73.**) Give the rule for reducing fractions to their least common denominator. (**74.**) What is addition of fractions? (**75.**) Give the rule for addition of fractions. (**76.**) What is subtraction of fractions? (**77.**) Give the rule for subtraction of fractions. (**78.**) What is multiplication of fractions? (**79.**) Give the rule for the multiplication of fractions. (**80.**) What is division of fractions? (**81.**) Give the rule for division of fractions. (**82.**) Give a rule for multiplying by 25. Give a rule for dividing by 25.

II. DECIMAL FRACTIONS.

DEFINITIONS.

83. A **Decimal Fraction** is a fraction whose denominator is 10, 100, 1,000, or some higher *power* of 10, (Art. **46**). Thus, $\frac{3}{10}$, $\frac{14}{100}$, $\frac{27}{1000}$, etc., are *decimal fractions*.

MENTAL EXERCISES.

1. If 1 is divided into 10 equal parts, what is one of the parts called? 2 of the parts? 5 of the parts?

2. If $\frac{1}{10}$ is divided into 10 equal parts, what is one of the parts called? 2 of the parts? 17 of the parts?

3. If $\frac{1}{100}$ is divided into 10 equal parts, what is 1 of the parts called? 3 of the parts? 27 of the parts?

4. What is $\frac{1}{10}$ of $\frac{1}{10}$? $\frac{1}{10}$ of $\frac{1}{100}$? $\frac{1}{10}$ of 1,000? $\frac{1}{10}$ of 10,000? What power of 10 is 100? 1,000? 10,000? 100,000? 1,000,000?

DECIMALS, AND THE DECIMAL POINT.

84. Decimal fractions may be written in two ways: **their denominators may be expressed,** as in ordinary fractions; or **their denominators may be indicated** by means of a **point** followed by one or more figures. In the latter case they are called **Decimals,** and the point (.) used in writing them is called the **Decimal Point.**

NOTATION OF DECIMALS.

85. Decimals are written in the same manner as whole numbers, and both may be written together, decimals on the *right* and whole numbers on the *left,* as shown in the following

NUMERATION TABLE.

etc., etc.	tens of millions.	millions.	hundreds of thousands.	tens of thousands.	thousands.	hundreds.	tens.	units.		tenths.	hundredths.	thousandths.	ten-thousandths.	hundred-thousandths.	millionths.	ten-millionths.	etc., etc., etc.
... 8	3	9	6	3	o	4	2	.	5	7	4	9	8	2	6 ...

Whole Numbers.   Decimals.

NOTE.—In whole numbers, *places of figures and orders of units* are counted from the decimal point toward the *left ;* in decimals, they are counted from the decimal point toward the *right.*

A figure in the **first place** of decimals denotes **tenths;** in the **second place** it denotes **hundredths;** in the **third place** it denotes **thousandths;** and so on, as indicated in the table. Hence, to write a decimal we have the following

RULE.

Write the number of tenths in the first decimal place, the number of hundredths in the second place, the number of thousandths in the third place, and so on.

EXAMPLES.

1. Write *three/tenths,* as a decimal. *Ans.* .3.
2. Write *twenty seven/hundredths.* *Ans.* .27.
3. Write *forty eight/thousandths.* *Ans.* .048.

NOTE.—In Example 2, because 27 *hundredths* is the same as 2 *tenths* and 7 *hundredths,* we write 2 in the *first* place of decimals and 7 in the *second* place. In Example 3, because 48 *thousandths* is the same as 0 *tenths,* 4 *hundredths,* and 8 *thousandths,* we write 0 in the *first* place, 4 in the *second* place, and 8 in the *third* place. Let the student in like manner explain each of the following examples:

4. *Two hundred and thirteen/thousandths.*

5. *One thousand and six/ten thousandths.*

6. *Four thousand two hundred and seven/millionths.*

7. *Two hundred and seventy four thousand three hundred and forty three/millionths.*

8. *Twenty three million, two hundred and four thousand, five hundred and seventy seven/hundred millionths.*

A mixed decimal is a mixed number whose fractional part is a *decimal.* Thus, six, and *three/tenths* is a mixed decimal; it may be written 6.3. In all such cases the integral part is written on the left of the decimal point.

9. Twenty, and *forty four/hundredths.* *Ans.* 20.44.

10. Thirty seven, and *seventy two/thousandths.*

11. Forty seven, and *two hundred and nine/millionths.*

NOTE.—In the preceding examples, decimals are in *italics.* The sign / separates the numerator from the denominator.

From what precedes, we see that a *decimal fraction* may be expressed **decimally** by writing its numerator, and then placing a decimal point so that the number of figures following it shall be equal to the number of ciphers in the denominator. If the number of figures in the numerator is less than the number of ciphers in the denominator, a sufficient number of ciphers must be **prefixed,** that is, *written before* the numerator.

EXAMPLES.

1. $\frac{3}{10} = .3.$ 4. $9\frac{311}{1000} = 9.311.$

2. $\frac{45}{1000} = .045.$ 5. $4\frac{79}{1000} = 4.079.$

3. $\frac{17}{10000} = .0017.$ 6. $256\frac{117}{100000} = 256.00117.$

NUMERATION OF DECIMALS.

86. From what has been explained, we see that a decimal may be read by the following

RULE.

Read the significant part as a whole number, and add the name of the lowest unit of the decimal.

NOTE.—Before reading a decimal the pupil should *numerate* it, that is, he should begin at the left hand and name the units of each place : thus, *tenths, hundredths, thousandths,* etc., according to the table.

Read the following decimals :

1. .087.   *Ans. Eighty seven|thousandths.*

2. .000317.

         Ans. Three hundred and seventeen|millionths.

3. .0027.	6. .52346.	9. .11122.
4. .10364.	7. .50067.	10. .224785.
5. .00201.	8. .320315.	11. .0067412.

NOTE.—In mixed decimals we read the integral and the decimal parts separately.

12. 120.009.

Ans. One hundred and twenty, and *nine/thousandths.*

13. 19.00015.   15. 150.15632.   17. 45.36251.

14. 212.1236.   16. 34.001725.   18. 111.009265.

DECIMAL CURRENCY.

87. The currency of the United States is purely decimal, the primary unit being 1 **Dollar.** In it **Dimes** are *tenths* of a dollar, **Cents** are *hundredths* of a dollar, and **Mills** are *thousandths* of a dollar. Dollars, cents, and mills are generally written in the form of a mixed decimal, the decimal point being placed after *dollars.* Thus, the expression $74.853, denotes 74 *dollars,* 8 *dimes,* 5 *cents,* and 3 *mills ;* it is read 75 *dollars* 85$\frac{3}{10}$ *cents.*

NOTE.—An *eagle* is equal to $10. In business transactions the terms *eagle, dime,* and *mill* are but little used, sums of money being expressed in *dollars* and *cents.*

FUNDAMENTAL PRINCIPLES.

88. Moving the decimal point one place to the right changes *tenths* to *units, hundredths* to *tenths,* and so on; but this is equivalent to multiplying the decimal by 10: hence, the following principle:

1°. *Moving the decimal point one place to the right is equivalent to multiplying the decimal by 10.*

In like manner we have the following principle:

2°. *Moving the decimal point one place to the left is equivalent to dividing the decimal by 10.*

Annexing a cipher to a decimal multiplies both numerator and denominator by 10; but this does not alter the value of the fraction (Art. **65**); hence, the following principle :

3°. *Annexing one or more ciphers to a decimal does not change its value.*

In like manner we have the following principle :

4°. *Striking out one or more terminal ciphers does not change the value of a decimal.*

REDUCTION OF COMMON FRACTIONS TO DECIMALS.

89. Let it be required to reduce, that is, to change $\frac{5}{8}$ to the form of a decimal.

EXPLANATION.—The value of $\frac{5}{8}$ is equal to $5 \div 8$ (Art. **62**); to find this quotient we annex three ciphers to 5, which is equivalent to multiplying it by 1,000, and then perform the division ; but this result is 1,000 times as great

OPERATION.

$$8)5000$$
$$.625$$

as the true value of the fraction ; we therefore divide it by 1,000, which is done by pointing off three decimal figures (Prin. 2, Art. **88**).

In like manner we may treat all similar cases ; hence, the

RULE.

Annex ciphers to the numerator and divide the result by the denominator ; then point off from the right of the quotient a number of decimal figures equal to the number of ciphers annexed.

NOTE.—If the number of figures in the quotient in less than the number of ciphers annexed, prefix the requisite number of ciphers.

EXAMPLES.

Reduce the following fractions to decimals :

1. $\frac{3}{4}$.

2. $\frac{15}{16}$.

3. $\frac{27}{64}$.

4. $\frac{118}{125}$.

5. $\frac{478}{625}$.

6. $\frac{849}{3125}$.

7. $\frac{127}{256}$.

8. $\frac{6}{125}$.

9. $\frac{3}{1250}$.

Note.—To reduce a mixed number to a decimal form, we reduce the fractional part to a decimal and annex the result to the integral part.

10. $19\frac{7}{8}$. 12. $11\frac{15}{128}$. 14. $21\frac{7}{3125}$.

11. $24\frac{13}{26}$. 13. $110\frac{4}{125}$. 15. $4\frac{149}{1600}$.

APPROXIMATE RESULTS.

90. It may happen that the division described in the last article will *not terminate,* no matter how many ciphers we annex. In this case the decimal found by stopping at any particular step of the division is called an **approximate** value of the given fraction. Thus .1904 is an *approximate* value of $\frac{4}{21}$. In this case $\frac{4}{21}$ is greater than .1904 and less than .1905; hence, it differs from either by less than they differ from each other, that is, by less than .0001. In like manner the approximate value of a fraction found by stopping at any decimal figure differs from the true value of the fraction by less than the corresponding decimal unit.

If we stop at any decimal figure and increase it by 1 when the next figure is equal to, or greater than 5, the error can never exceed $\frac{1}{2}$ the corresponding decimal unit. Thus, $\frac{2}{3} = .667$, and $\frac{1}{3} = .333$, each to within less than $\frac{1}{2}$ of .001. This is the **practical method** of finding approximate values of decimals.

Note.—In applying the principles of decimals to practical cases we shall habitually follow the method of approximation just explained, and, except in special cases, we shall limit the approximation either to three or to four decimal places.

In United States money we shall habitually limit the approximation to three decimal places; each result will then be true to within *half a mill,* or the *twentieth of a cent.*

Reduce the following fractions to decimals, carrying the approximation to the fourth place:

1. $\frac{3}{4}$ of $\frac{1}{4}$.

2. $\frac{17}{24}$.

3. $\frac{14}{21}$.

4. $\frac{4}{31}$.

5. $3\frac{14}{17}$.

6. $4\frac{10}{11}$.

7. $\frac{1}{2}$ of $3\frac{1}{4}$.

8. $\frac{3}{11}$ of $7\frac{1}{3}$.

9. $\frac{2}{3}$ of $\frac{5}{8}$ of $4\frac{1}{8}$.

10. $7\frac{1}{2} \times 4\frac{3}{4}$.

11. $21\frac{31}{32}$.

12. $1\frac{22}{225}$.

13. $14\frac{14}{18}$.

14. $13\frac{11}{117}$.

15. $4\frac{18}{525}$.

ADDITION OF DECIMALS.

DEFINITION.

91. Addition of Decimals is the operation of finding the **sum** of two or more decimals.

MENTAL EXERCISES.

1. What is the sum of 4 *tenths* and 5 *tenths?* of .3 and .6 ? How many *units* and *tenths of a unit* in the sum of .8 and .9 ? What is the sum of .3, .5, and .9? of .5, .7, .9, and .6 ?

2. How would you read 49 *hundredths* in *tenths* and *hundredths?* How many *hundredths* are there in 6 *tenths?* What is the sum of .5 and .49 ? of .59 and .4 ? How would you read *three hundred and seventeen/thousandths* in *tenths, hundredths*, and *thousandths?*

3. What is the sum of 13 *cts.* and 22 *cts.?* of $.13 and $.22 ? Is there any difference between 13 *cts.* and $.13 ? What is the sum of 25 *cts.* and $.45 ? of $.4, $.25, and $.7 ? of .4, .25, and .7 ?

Note.—*Addition of decimals* depends on the same principles as *addition of integers.*

OPERATION OF ADDITION.

92. Let it be required to find the sum of 4.035, 76.19, and 114.0305.

EXPLANATION.— We write the decimals so that units of the same order shall stand in the same column ; this will bring all the decimal points in one column: then beginning at the right, we add each column separately, *setting down* and *carrying* as in simple numbers. Hence, the

OPERATION.

$$4.035$$
$$76.19$$
$$114.0305$$
Sum. 194.2555

RULE.

Write the decimals so that units of the same order shall stand in the same column, and add as in simple numbers.

NOTE.—The decimal points of the numbers to be added, and of their sum, must stand in the same column.

EXAMPLES.

(1.)	(2.)	(3.)	(4.)	(5.)
3.057	5.6000	5.43	0.105	$3.97
14.086	17.0032	12.998	0.0012	$4.295
209.3154	35.9070	317.0971	0.25	$11.464
226.4584	58.5102	335.5251	0.3562	$19.729

6. Find the sum of .632, .718, 3.202, and 111.1.

7. Of .0049, 47.0426, 37.041, and 360.0039.

8. Of $81.053, $67.412, $93.172, and $14.38.

9. Of $59.317, $69.565, $8.213, and $7.775.

10. Of 3.25 *lbs.*, 47.348 *lbs.*, 748.4 *lbs.*, and 29.32 *lbs.*

11. Of 672.5 *yds.*, 4.923 *yds.*, 80 *yds.*, and .0764 *yds.*

12. Of 72.5 + 140 + 340.03 + 21.5715 + 4.0008.

13. Of 2.8146 + .0938 + 8.875 + 231.2788 + 4.0087.

14. Of 54.3 *ft.* + 7.29 *ft.* + 180.0046 *ft.* + 187 *ft.* + 3.024 *ft.*

15. Of 57.038 + 95.00487 + 53.4690 + 107.00003.

16. Of $62.70 + $2.03 + $4.009 + $78.15 + $114.

17. Of .0009 + 3.0021 + .128 + 8.0469 + 59.

18. Of 3.0102 *bu.* + 11.5008 *bu.* + 73.07 *bu.* + 2.92 *bu.* + 9.5 *bu.*

19. 2.005 + 110.301 + .069 + 7.375 + 2.25 = ?

20. 17.215 + 3.0567 + 2.072 + 4.009 + 54.75 = ?

21. 29.157 *ft.* + 8.0016 *ft.* + 77.29 *ft.* + 32.004 *ft.* + 8.848 *ft.* = ?

22. 14.2351 + 651.012 + 2.219 + 3.157 + 13.614 = ?

23. $861.55 + $378.25 + $461.37 + $683.57 + $1,205.47 = ?

24. 213.7 *bu.* + 2.913 *bu.* + 14.769 *bu.* + .0078 *bu.* = ?

25. 15.753 *yds.* + 2.069 *yds.* + 17.6143 *yds.* + 10.27 *yds.* + 3.2107 *yds.* = ?

PRACTICAL PROBLEMS.

1. A boy paid 28 *cts.* for a slate, 75 *cts.* for paper, and 94 *cts.* for an Arithmetic; what did he pay for all?

Ans. $1.97.

2. One field contains 5.3 *acres,* a second contains 11.43 *acres,* a third contains 17.59 *acres,* and a fourth contains 3.175 *acres*; how many acres in all of them?

3. A. bought 16 hams for $31.87½, a bag of coffee for $17.92, a chest of tea for $12.75, and a firkin of butter for $21.37½; what did they all cost?

4. A man bought candles for $6.89, flour for $25.56, raisins for $1.12½ (*i. e.* for $1.125), cheese for $8.37½, and sugar for $5.44; what was the cost of the whole?

5. A merchant sold 4 pieces of muslin; the first contains 34.25 *yds.*, the second 38.056 *yds.*, the third 40.2 *yds.*, and the fourth 37,225 *yds.*; how many yards in all?

6. A farmer has 4 bins of wheat; in the first there are 86.35 *bu.*, in the second 73.125 *bu.*, in the third 96.5 *bu* , and in the fourth 74.3 *bu.*; how many bushels in all?

7. In 5 piles of wood there are respectively 4.316 *cords*, 8.23 *cords*, 11.25 *cords*, 7.364 *cords*, and 13.819 *cords;* how many *cords* are there in all the piles?

8. B. bought a house for $5,000, a store for $6,290, merchandise for $23,654.12, a horse for $278.53, a farm for $9,371.60, bank stock for $11,500, and a watch for $92.72½; what did the whole cost him?

SUBTRACTION OF DECIMALS.

DEFINITION.

93. Subtraction of Decimals is the operation of finding the **difference** between two decimals.

MENTAL EXERCISES.

If 6 *tenths* are taken from 9 *tenths*, how much will remain? What is the difference between 23 *tenths* and 14 *tenths?* 23 *tenths* is equal to how many *units*, and how many *tenths?*

2. What is the difference between 35 *hundredths* and 2 *tenths?* How many *tenths* in 35 *hundredths?* What is the difference between .45 and .25? 4.5 and .6? .7 and .15? 4.5 and 2.2? 3.5 and 2.7? 3.2 and 1.9?

NOTE.—*Subtraction of decimals* depends on the same principles as *subtraction of integers.*

OPERATION OF SUBTRACTION.

94. Let it be required to subtract 4.079 from 11.362.

EXPLANATION.—The subtrahend is written under the minuend, so that units of the same order shall stand in the same column; this will bring the decimal points into the same column: the operation is then performed as in the subtraction of simple numbers. Hence, the following

OPERATION.

Minuend,	11.362
Subtrahend,	4.079
Remainder,	7.283

RULE.

Write the subtrahend under the minuend, so that units of the same order shall stand in the same column; then subtract as in simple numbers.

NOTE.—The decimal points of the *minuend*, of the *subtrahend*, and of the *remainder* must stand in the same column.

EXAMPLES.

	(1.)	(2.)	(3.)	(4.)
Minuend,	5.316	17.0091	1075.0567	$312.475
Subtrahend,	2.013	11.9902	287.9374	$214.268
Remainder,	3.303	5.0189	787.1193	$98.207

NOTE.—If the subtrahend contains more decimal figures than the minuend, annex the requisite number of ciphers to the minuend, or conceive them to be annexed (Principle 3°, Art. **88**).

(5.)	(6.)	(7.)	(8.)	(9.)
13.700	13.7	884.1300	884.13	$8.
8.299	8.299	33.7865	33.7865	$4.735
5.401	5.401	850.3435	850.3435	$3.265

10. From 298.789 subtract 196.493.   *Ans.* 102.296.

11. From 2684.11 subtract 199.8637.   *Ans.* 2484.2463.

Find the difference between,

12. 127.334 and 55.827.

13. 94.8607 and 27.861.

14. 986.444 and 98.6438.

15. $17.025 and $7.255.

16. 2.867 *ft.* and .9965 *ft.*

17. $661.40 and $95.472.

18. $25,000 and $1.077.

19. 100 *yds.* and 99.001 *yds.*

20. 41.02 and 40.021.

21. 35 *ft.* and 25.0003 *ft.*

22. 6 *yds.* and .0006 *yds.*

23. $14.003 and $9.875.

24. 13.4072 and 9.1875.

25. 18.65 and 12.0734.

26. 17.314 and 12.9921.

27. 13.3125 and 8.4139.

28. $34.883 and $9.43.

29. 87.007 *lbs.* and 10.895 *lbs.*

30. $1.87 + $3.945 + $27 − ($6.42 + $15.07 + $.25) = ?

31. 125.6 *lbs.* + 27.42 *lbs.* − (4.3 *lbs.* + 12.11 *lbs.* + 9 *lbs.*) = ?

32. $5,000 + $325.175 − ($2,710.75 − $147.56) = ?

33. ($794.26 − $215.875) − ($456.375 − $211.12) = ?

PRACTICAL PROBLEMS.

1. Mr. Holmes bought a cow for $45.125 and sold her for $49.18; what did he gain?

2. From a piece of cloth containing 42.37 *yds.*, 16.89 *yds.* were cut off; how many yards remained in the piece?

3. What is the difference between $875.043 and $704.91?

4. How much must I add to $617.37½ to make $922.75?

5. A.'s income is $6,250 per year, of which he spends $3,142.75 and lays up the rest; what does he lay up?

6. From $981.43 + $456.81 subtract $498.75.

7. From $10,000 subtract $4,367.18 + $3,587.47.

8. From $965 + $341.60 subtract $433.33 + $89.47.

9. A man received the following sums: $27.40, $68.75, $810.47, $386.59, and $2.20; he paid out the following sums: $78.67, $129.72, $119.46, and $3.88; how much had he left?

10. A. had, at the beginning of the year, goods worth $10,500; during the year he bought goods to the amount of $9,345.75, and sold to the amount of $13,450.95; at the close of the year he had goods worth $11,122.37; how much did he make during the year?

11. A lady bought a dress for $42.18, a bonnet for $17.65, and a pair of gloves for $1.87½; she gave for them a $100 bill; how much change ought she to receive?

12. A. bought 37.41 cords of wood, of which he sold 8.3 C. and burned 13.426 C.; how many cords had he left?

13. A flagstaff is made up of two parts, the upper part being 27.84 ft. long, and the lower part 57.86 ft. long: now if the lower part is set 11.31 ft. in the ground, how many feet of the whole staff is above the ground?

14. A man had $137.26, of which he spent $17.87½ for coal, $22.12½ for flour, $7.42 for soap, and $32.79 for a suit of clothes; how much had he left?

15. From a hogshead of sugar containing 397.25 lbs., a grocer sold parcels as follows: 110.25 lbs., 64.5 lbs., 14.25 lbs., 29.375 lbs., 39.23 lbs., and 16.33 lbs.; how much was left?

16. A. is to travel 597½ miles in 3 days; the first day he travels 196.4 miles, and the second day he travels 201.25 miles: how many miles must he travel the third day?

17. A farmer had a colt worth $147½, which he traded for a cow worth $42.375, 4 calves worth $22¾, and the balance in cash; how much cash did he receive?

18. A merchant bought a piece of cloth for $75¼, a box of ribbons for $25⅜, and a quantity of thread for $27.87; he sold the cloth for $87.125, the ribbons for $22.16, and the thread for $21⅜: did he gain or lose, and how much?

MULTIPLICATION OF DECIMALS.

DEFINITION.

95. Multiplication of Decimals is the operation of finding the **product** of two decimals.

MENTAL EXERCISES.

1. If a copy-book costs 2 *tenths* of a dollar, what will 4 copy-books cost? What is 4 times 2 *tenths* of a dollar? What is 6 times 2 *tenths?* How many *units* and how many *tenths* in the product? How many tenths are 6 times 9 *tenths?* How many *units* and how many *tenths* in the product?

2. If a melon costs 3 *tenths* of a dollar, what does 1 *tenth* of a melon cost? What is 1 *tenth* of 3 *tenths?* What is the product of .3 by .1 ? of .3 by .2 ? of .3 by .5 ? of .7 by .9? of .8 by .8 ? What is the decimal unit of the product of *tenths* by *tenths?*

3. What is 1 *tenth* of 4 *hundredths?* What is the product of 4 *hundredths* by 1 *tenth?* of .04 by .2 ? of .04 by .4 ? of .05 by .7 ? of .09 by .9 ? What is the decimal unit of the product of *hundredths* by *tenths?*

4. What is the product of .005 by .2 ? of .007 by .09 ? of .006 by .008? of .7 by .6 ? of .7 by .04 ? of .03 by .07 ?

5. What is *one hundredth* of *one hundredth?* What is the product of 3 *hundredths* by 7 *hundredths?* What is the decimal unit of the product of *hundredths* by *hundredths?* What is the product of .09 by .06? of .07 by .08 ?

NOTE.—*Multiplication of decimals* depends on the same principles as *multiplication of integers,* and also on the rule for the *multiplication of common fractions.,* (Art. **79**).

OPERATION OF MULTIPLICATION.

96. Let it be required to find the product of 7.8 and .82.

EXPLANATION.—Here the decimals are first changed to equivalent common fractions and multiplied together by the rule of Article 79 ; the resulting fraction is then reduced to the decimal form ; in doing this, we have actually multiplied the given decimals together, without reference to their decimal points, and in the result we have pointed off as many decimal figures as there are in both factors.

OPERATION.

$$7.8 \times .82 = \frac{78}{10} \times \frac{82}{100}$$

$$= \frac{6396}{1000} = 6.396.$$

Since all similar cases may be treated in the same manner, we have the following

R U L E .

Multiply as in simple numbers, and point off, from the right of the product, as many decimal figures as there are in both factors.

NOTE.—If the number of places in the product is less than the number of decimal places in both factors, prefix as many ciphers as may be necessary.

E X A M P L E S .

1. What is the product of 3.05 by 4.102 ?

Ans. 12.5111.

2. What is the product of .003 by .042 ?

Ans. .000126.

Perform the following multiplications :

3. $38.4 by 16.7.

4. $14.25 by .375.

5. 1,500 *bu.* by .00014.

6. $1.009 by .0012.

7. 146.05 *yds.* by 128.6.

8. .565 *ft.* by .16.

9. .0463 *lbs.* by .0081.

10. 701.005 × 60.06.

11. 456.05 × 3.825.

12. 308.25 × .0775.

13. 27.032 × 14.3.

14. $380.06 × 22.

15. $24.07 × .125.

16. $.75 × .33.

17. $456.87 × .066.

18. $798.007 × .08.

19. $.034 × .08.

20. 7.45 *lbs.* × 2.7504.

21. 42.2 × 2.004.

22. 79.004 × .00473.

23. 412.5384 × 1.00003.

24. 40.86 *yds.* × .00293.

25. 0.0756 *rods* × 6.75.

26. .2897 × 3020.

27. $37.55 × 45.64.

28. 3.005 × 21.82 × 14.71.

29. 8.013 × 11.7 × 0.774.

30. 12.12 × 300.7 × 8.004.

31. 0.713 × 2.346 × 2.005.

32. $12.5 × 7.2 × 16.5.

33. 4.2 *lbs.* × 8.1 × 2.4.

34. 1.7 *yds.* × 11.4 × 82.3.

NOTE.—To multiply a decimal or a mixed decimal by 10, 100, 1,000, etc., move the decimal point as many places to the *right* as there are ciphers in the multiplier, annexing ciphers to the multiplicand if necessary.

35. What is the product of 77.56 by 10 ?   *Ans.* 775.6.

36. What is the product of .0075 by 100 ?   *Ans.* .75.

37. What is the product of 6.6 by 1000 ?   *Ans.* 6600.

38. ($31.45 + $18.2) × 7.2 — $240.15 = ?

39. (180.6 *lbs.* — 36.4 *lbs.*) × (67.2 — 3.47) = ?

40. $150.75 × 16.3 + $211.5 × 16 — $114.25 × 9 = ?

41. (38.4 *yds.* + 56.4 *yds.*) × 7.2 — 18.36 *yds.* × 8.1 = ?

42. (463.45 + 31.4 — 2.175) × (18.2 — 11.07) = ?

PRACTICAL PROBLEMS.

1. What is the cost of 17 barrels of flour at $6.37½ a barrel ?

2. Of 85½ *lbs.* of tea at $1.37½ a pound ?

3. Of 311 *yds.* of linen at 64½ *cts.* a yard ?

4. Of 41½ gallons of wine at $3.12½ a gallon ?

5. Of 278 cords of wood at $9.62½ a cord ?

6. Of 17 *lbs.* of tea at $.75 a pound ?

7. Of 7.5 reams of paper at $3.62½ a ream ?

8. Of 2,754 sheep at $5.12½ apiece ?

9. Of 47.75 *bu.* of corn at $.875 a bushel ?

NOTE.—Let the student apply the rule for approximate results explained in Art. **90,** finding values to the nearest mill. To secure uniformity the rule should be applied at each step of the operation.

10. A grocer sold 25.5 *lbs.* of sugar at 12½ *cts.* a pound, and 18.6 *lbs.* of lard at 13½ *cts.* a pound; how much did he receive for both ?

11. A farmer sold 37½ *bu.* of oats at 42½ *cts.* a bushel, and 35¼ *bu.* of potatoes at 37½ *cts.* a bushel; he received for the same 43¼ *yds.* of muslin at 12½ *cts.* a yard, and the balance in cash: how much cash did he receive ?

12. A. sold 75 *bu.* of wheat at $1.12½ a bushel, 36.2 *bu.* of beans at $2.37½ a bushel, and 97¼ *lbs.* of butter at 22½ *cts.* a pound; what did he receive for the whole ?

13. A man's wages are $18.87½ a week, and his expenses are $13.25 a week; how much can he save in 14¼ weeks ?

14. A man was to walk 245⅔ in 7 days: for the first 3 days he walked at the rate of 34.36 miles a day, and for the next three days he walked 36.75 miles a day; how far had he to walk the seventh day ?

15. The distance from St. Louis to New Orleans is 1332 miles; two boats start at the same time, one from St. Louis, and the other from New Orleans, are run towards each other; the boat from St. Louis makes 230⅔ miles a day, and the one from New Orleans 196⅔ miles a day: how far apart are they at the end of 2½ days ?

16. A man starts from a certain point and travels in a certain direction at the rate of 7.25 miles an hour; at the

end of 2½ hours a second man starts from the same point and travels in an opposite direction at the rate of 6.29 miles an hour: how far apart are they at the end of the sixth hour?

17. A carpenter earned $12.87½ a week for 3 weeks; the first week he spent $8.333, the second week he spent $9.18, the third week he spent $7⅜, and the rest he saved; how much did he save?

18. A gardener sold his cabbages for $212.87½, and his turnips for $118.33; the cost of raising the cabbages was $119.75, and the cost of raising the turnips was $99.87½: what was his profit on the two crops?

19. A man bought 43 sheep at the rate of 4 *dollars* and 67½ *cents* a piece, and sold the lot for 215 *dollars* and 42¼ *cents;* did he gain or lose, and how much?

20. A man made a journey as follows: he traveled 7¾ hours by rail at the rate of 22.75 miles an hour, 9¼ hours by stage at the rate of 6.75 miles an hour, and 11.75 hours on foot at the rate of 4.62 miles an hour; what was the length of the journey?

DIVISION OF DECIMALLS.

DEFINITION.

97. Division of Decimals is the operation of finding the **quotient** of one decimal by another.

MENTAL EXERCISES.

1. What is the product of .3 by .5 ? What then is the quotient of .15 by .5 ? How many decimal places in the *dividend?* in the *divisor?* in the *quotient?*

2. What is the product of .12 by .13? What then is

the quotient of .0156 by .12 ? by .13 ? How many decimal places in the *dividend?* in the *divisor?* in the *quotient ?*

3. What is the product of .003 by .9 ? What then is the quotient of .0027 by .9 ? How does the number of decimal places in the *quotient* compare with the number in the *dividend* and in the *divisor.*

NOTE.—*Division of decimals* depends on the same principles as division of integers, and also on the rule for the division of fractions (Art. **81**).

OPERATION OF DIVISION.

98. Let it be required to divide 7.8 by .125.

EXPLANATION.—Here we have reduced the given decimals to common fractions and divided by the rule of Art. **81**; we have then reduced the result to the decimal form; in doing this, we have actually annexed three decimal ciphers to the dividend (Art. **88**), and divided the result by the divisor, without reference to the decimal point; then from the quotient we have pointed off as many decimal figures as the number in the reduced dividend exceeds that in the divisor.

OPERATION.

$$\frac{7.8}{.125} = \frac{78}{10} \div \frac{125}{1000}$$

$$= \frac{78}{10} \times \frac{1000}{125} = \frac{78000}{125} \div 10$$

$$= 624 \div 10 = 62.4$$

All similar cases may be treated in the same manner; hence, the following

R U L E .

Annex decimal ciphers to the dividend if necessary; then divide as in simple numbers and point off from the right of the quotient as many decimal figures as the number of decimal places in the dividend exceeds that in the divisor.

5

Notes.—1. The dividend *must* contain *as many* decimal figures as the divisor, but it *may* contain *more*. If the number of decimal figures is the same in both, the quotient is a whole number.

2. If the number of figures in the quotient is less than that required by the rule, a sufficient number of ciphers must be prefixed.

EXAMPLES.

1. Divide 40.05 by 4.5.   *Ans.* 8.9.

2. Divide .0141 by .00047.   *Ans.* 30.

3. Divide 2.3 by 1437.5.   *Ans.* .0016.

Perform the following indicated divisions, limiting approximate values to the fourth decimal place:

4. .00125 ÷ .5.

5. $34.75 ÷ 25.

6. 46.103 ÷ 2.14.

7. 7.8125 ÷ 31.25.

8. $2756.25 ÷ 31.5.

9. $68.875 ÷ 14.5.

10. 3414.52 ÷ 30.25.

11. 16.025 ÷ .045.

12. .9375 *ft.* ÷ .075.

13. 112.1184 ÷ 9.16.

14. 9322.15 ÷ 6.275.

15. $5.875 ÷ 35.25.

16. 480 ÷ 3.12.

17. $1.8 ÷ 28.8.

18. 17.1031 *yds.* ÷ .53.

19. .09925 ÷ .37.

20. 3.72812 ÷ 4.07.

21. $18.1771 ÷ 27.13.

22. 101.6688 ÷ 43.08.

23. 1.51088 ÷ .019.

24. 187.12264 ÷ 1.52.

25. $71.1022 ÷ $9.43.

Note.—To divide a decimal or a mixed decimal by 10, 100, 1000, &c., move the decimal point as many places to the *left* as there are ciphers in the divisor, prefixing ciphers to the dividend if necessary.

26. What is the quotient of 77.56 by 10?   *Ans.* 7.756.

27. What is the quotient of .0075 by 100? *Ans.* .000075.

28. What is the quotient of 6.6 by 1000?   *Ans.* .0066.

29. ($28 + $11.75) ÷ 1.25 + $38.75 = ?

30. $50 ÷ 5.75 + ($10 − $3.75) × 1.2 = ?

31. ($13.75 − $1.87½) ÷ (12.75 − 4.5) = ?

32. $(63.5\,ft. - 24.25\,ft.) \div (17.25 - 11.75) = ?$

33. $4\frac{1}{2}\,yds. \times 2.2 + 7\frac{1}{2}\,yds. \div 1.25 + 18.875\,yds. = ?$

34. $(47.3\,lbs. + 6.7\,lbs.) \div (34.18 - 16.78) = ?$

PRACTICAL PROBLEMS.

1. If a man can earn $519.75 in 13.5 weeks, how much can he earn in 1 week ?   *Ans.* $38.50.

2. If 20.5 *bu.* of buckwheat cost $12.71, what is the cost of 1 *bu.*? of 7½ *bu.*?   *Ans.* 62 *cts.*; $4.65.

3. If a barrel of flour costs $5.75, how many barrels can be bought for $1,035 ?   *Ans.* 180 *bbls.*

4. If 1,000 *acres* of land cost $17,586, what is the cost of 1 *A.*? of 75½ *A.*?   *Ans.* 17.58\frac{6}{10}$; $1,32.743.

5. If 1 *acre* costs $25.62, how many acres can be bought for $1,242.57 ?

6. If 75.3 cords of wood cost $640.05, how much will 1 *C.* cost ? 6.3 *C.*?

7. If 1 *cord* of wood costs $8.25, how much wood can be bought for $156.75 ?   For $30.52½ ?

8. At $4.28 a yard, how much cloth can be bought for $74.90 ?   How much for 152.52\frac{4}{10}$?

9. A farmer sold 27.5 *lbs.* of butter at 20 *cts.* a pound, and 17.5 *bu.* of oats at 75 *cts.* a bushel; he took in payment in sugar at 12½ *cts.* a pound: how many pounds did he receive ?

10. There are 31.5 gallons in a barrel; how many barrels are there in 2756.25 gallons ?

11. There are 1,760 *yds.* in 1 mile; how many miles are there in 23,760 *yds.?*   In 26,840 *yds.?*

12. A merchant buys a piece of cloth containing 35 *yds.*

for $87.50; he wishes to sell it so as to gain $17.50: at what price must he sell it per yard?

13. A man can travel 441.5 *miles* in 7.5 days; how far can he travel in 1 day? in 9½ days?

14. A man travels 7.25 days at the rate of 211.5 miles a day; on his return he makes the whole journey in 5 days: how many miles is that per day?

15. A. and B. start at the same time from points 147.16 *miles* apart, and travel toward each other till they meet; if A. travels at the rate of 7.75 *mi.* an hour, and B. at 6.4 *mi.* an hour, how long before they meet?

16. A speculator bought 78.25 acres of land for $9,781.25, and sold it so as to gain $3.50 an acre; what did he get per acre?

17. If 527 *bu.* of wheat cost $592.87½, what is the cost of 1 *bu.?* of 6.5 *bu.?*

18. If 115 *lbs.* of beef cost $19.89½, what will 93 *lbs.* cost at the same rate?

19. If I pay $39.48 for 28 *bu.* of wheat, what must I pay for 48 bushels at the same rate?

20. A grocer bought 114 gallons of vinegar at 22½ *cts.* a gallon, and sold it so as to gain $7.98; at what rate per gallon did he sell it?

21. A. starts from a certain place and travels along a road at the rate of 4.66 miles an hour; B. starts 13.75 miles behind him and travels in the same direction at the rate of 5.91 miles an hour: how long before B. will overtake A.?

22. A farmer sold 22.5 *bu.* of wheat at $1.18 a bushel, and a certain number of bushels of oats at 68 *cts.* a bushel;

he received for his oats $22.41 more than he did for his wheat: how many bushels of oats did he sell?

23. If 11½ *lbs.* of coffee cost $2.07, what will 1 *lb.* cost? What will 31¼ *lbs.* cost?

24. How many bushels of oats at 62½ *cts.* a bushel will pay for 4¼ *thousand* of lumber at $7.50 a thousand?

25. A farmer exchanged 70 *bu.* of rye at $0.92 a bushel, for 40 *bu.* of wheat at $1.37½ a bushel, and the balance in oats at $0.40 a bushel; how many bushels of oats did he receive?

26. If a man can travel 32.48 *miles* in .8 of a day, how far can he travel in 5.3 days?

27. What is the sum of the quotients of 24 by 9.6, of 42.75 by 11.4, and of 17.85 by 4.2?

28. A. and B. start together and travel in the same direction around an island whose circuit is 4.2 miles; A. travels at the rate of 4.6 *mi.* an hour, and B. at the rate of 5.2 *mi.* an hour: how many hours before they are together again?

29. A man bought a farm containing 64.5 acres for $1,773.75; what was that per acre?

30. How many cords of wood at $8 a cord must be paid for 24 yards of cloth at $3.50 a yard?

31. If 60 bushels of turnips cost $18.60, how much will 19 bushels cost?

32. If 10 tons of coal cost $57.50, how many tons can be bought for $235.75?

33. A tailor cuts from 31.25 *yds.* of cloth 6 coats, each taking 3.75 *yds.*, and makes the rest into vests, each taking 1.25 *yds.*; how many vests does he make?

III. CONTRACTIONS AND BUSINESS OPERATIONS.

ALIQUOT PARTS.

99. An **Aliquot Part** of a number is one of the equal parts, whether integral or fractional, into which the number can be divided.

The principal aliquot parts of a dollar are shown in the following

TABLE.

50 cts., equal to $\frac{1}{2}$ of \$1.	12$\frac{1}{2}$ cts., equal to $\frac{1}{8}$ of \$1.
33$\frac{1}{3}$ cts.,　"　"　$\frac{1}{3}$ "　"	10 cts.,　"　"　$\frac{1}{10}$ "　"
25 cts.,　"　"　$\frac{1}{4}$ "　"	6$\frac{1}{4}$ cts.,　"　"　$\frac{1}{16}$ "　"
20 cts.,　"　"　$\frac{1}{5}$ "　"	5 cts.,　"　"　$\frac{1}{20}$ "　"

100. To find the cost of any number of things when 1 thing costs an **aliquot part of a dollar**, we have the following

RULE.

Divide the number of things by the number of times the price of one thing is contained in $1; the quotient will be the required number of dollars.

EXAMPLES.

1. What is the cost of 64 bushels of oats at 50 cents a bushel?　　　　　　　　*Ans.* \$$\frac{64}{2}$ = \$32.

2. What will 116 pounds of beef cost at 20 cents a pound?　　　　　　　　*Ans.* \$$\frac{116}{5}$ = \$23.20.

3. Of 250 melons at 25 *cts.* each ?

4. Of 144 pencils at 12$\frac{1}{2}$ *cts.* each ?

5. Of 75 oranges at 5 *cts.* each ?

6. Of 69 *yds.* of sheeting at 33⅓ *cts.* a yard ?

7. Of 50 dozen, marbles at 6¼ *cts.* a dozen ?

8. Of 73 *lbs.* sugar at 12½ *cts.* a pound ?

9. 17 *dozen* eggs at 25 *cts.* a dozen ?

10. Of 117 *quarts* berries at 20 *cts.* a quart ?

11. Of 47 *lbs.* coffee at 33⅓ *cts.* a pound ?

12. Of 145 *lbs.* rice at 6¼ *cts.* a pound ?

13. Of 87.3 *lbs.* coffee at 33⅓ *cts.* a pound ?

14. Of 315 *lbs.* sugar at 12½ *cts.* a pound ?

15. Of 70 *lbs.* of butter at 33⅓ *cts.* a pound ?

16. Of 35 *doz.* eggs at 20 *cts.* a dozen ?

101. To find the cost of things sold by the **hundred,** or by the **thousand,** we have the following

RULE.

Multiply the cost of 100, or 1000 things, by the number of things, and move the decimal point 2, or 3 places to the left.

EXAMPLES.

1. What is the cost of 460 oranges at \$3.50 *per* hundred ?

$$\textit{Ans. } \frac{\$3.50 \times 460}{100} = \$16.10.$$

2. Of 1,726 *ft.* of boards at \$3 *per* 1,000 *ft.* ?

3. Of 47,555 bricks at \$7.50 *per* 1,000 ?

4. Of freight on 8,714 *lbs.* at 62½ *cts,* a *hundred?*

5. Of 83,750 *ft.* of stone at \$60 a *thousand?*

6. Of 763 *lbs.* of pork at \$4.50 a *hundred?*

7. Of 511 *lbs.* of beef at \$7 a *hundred?*

8. Of 1,432 *lbs.* of pork at \$8.25 per *hundred?*

9. Of 8,741 *ft.* of plank at $30 per *thousand ?*

10. Of 4,875 *ft.* of boards at $17 per *thousand ?*

11. Of 7,320 papers tacks at $30 a *thousand ?*

12. Of 756 *ft.* of stone at $5 a *hundred ?*

13. Of 3,450 oysters at $2 a *hundred ?*

14. Of 7,846 *lbs.* of hay at 90 *cts.* a *hundred ?*

102. To find the cost of things sold by the **ton**, that is, by the **two thousand pounds,** we have the following

R U L E .

Multiply half the cost of a ton by the number of pounds, and move the decimal point in the product three places to the left.

EXAMPLES.

1. What is the cost of 3,475 *lbs.* of plaster at $7.50 per ton ?   *Ans.* $\dfrac{\$3.75 \times 3475}{1000} = \$13.03\frac{1}{8}$.

2. Of transporting 6,742 *lbs.* at $7 a ton ?

3. Of 6,527 *lbs.* of oats at $30½ a ton ?

4. Of 18,747 *lbs.* of coal at $8½ per ton ?

5. Of 8,142 *lbs.* of iron at $100 a ton ?

6. Of 3,120 *lbs.* of wool at $660 a ton ?

7. Of 1,620 *lbs.* of hay at $16 per ton ?

8. Of 5,782 *lbs.* of pig iron at $22 a ton ?

9. Of 7,711 *lbs.* of straw at $15 per ton ?

10. Of 8,824 *lbs.* of hay at $15 per ton ?

11. Of 3,509 *lbs.* of wheat at $40 a ton ?

12. Of 2,250 *lbs.* of coal at $7.40 a ton ?

13. Of 14,710 *lbs.* of crushed stone at $3 a ton ?

14. Of 16,318 *lbs.* of meal at $37 a ton ?

BILLS AND ACCOUNTS.

103. A bill is a written statement of *goods sold, services rendered,* or *money paid,* with the date of each *item.*

The ordinary form of a bill of items is shown below, in which @ stands for **at.**

Bridgeport, Ct., July 13, 1877.

Mr. G. S. Tompkins,

Bought of J. Madden & Co.

16 pounds of tea	@ $0.85	$15	60
27 " coffee . . .	@ .25	6	75
17 gallons of molasses .	@ .92	15	64
112 pounds of sugar . . .	@ .14	15	68
	Amount,	$51	67

Received payment,

J. Madden & Co

NOTE.—The party that *owes* money is called a debtor (Dr.), and the party to whom *money is due* is called a creditor (Cr.). When a bill is paid, it is receipted by writing the name of the creditor, or his authorized agent, at the bottom of the bill after the words **received payment.**

To find the **amount,** or **footing** of a bill:

Find the amount of each item separately, and then find the sum of the results.

EXAMPLES.

Find the footings of the following bills:

(1.) *St. Louis, Mo.*, July 15, 1877.

P. G. BISSEL,

Bought of L. SMITH.

2462 feet of hemlock boards . . . @	$7 per 1000.	
5410 " " " . . . @	$10 per 1000.	
600 " scantling . . . . . @	$11.75 per 1000.	
1012 " plank . . . . . . @	$1.25 per 100.	
77 " hewn timber . . . . @ 15 cts. per foot.		

Amt.,

Received payment.

(2.) *Columbus, O.*, Aug. 4, 1877.

J. D. SMITH,

Bought of A. HINE.

32 yds. silk . . . . @ $2.12½	
18 yds. alpaca . . . . @ 87½ cts.	
16 yds. chintz . . . . @ 24 cts.	
42 yds. muslin . . . . @ 21 cts.	
15 pieces tape . . . . @ 14 cts.	
3 pairs gloves . . . . @ $1.94	
12 pairs stockings . . . . @ 62½ cts.	

Amt.,

Rec'd payment.

(3.) *Cairo, Ill.*, Sept. 5, 1877.

S. L. MORSE,

To J. BRISTOW, *Dr.*

To labor, self, 4 days . . . . @ $4.	
" " man, 6 days . . . . @ $3.	
" materials	$17

Amount,

Received payment.

Write the following in proper form and find the foot-ings:

4. John Duffie bought of D. Plant, April 7, 1877, 17 *yds.* of calico @ 12½ *cts.*, 12 *yds.* of muslin @ 17 *cts.*, 2½ *yds.* of linen @ 73 *cts.*, and 9 spools of thread at 7 *cts.*; what was the amount of the bill?

5. Mrs. Churchill bought of Knapp & Co., July 29, 1877, the following items: 1 shawl @ $25.50, 22 *yds.* silk @ $2.25, 12 *yds.* lace @ 82 *cts.*, 3 *prs.* gloves @ $1.25, and 4 pieces tape @ 32 *cts.*: what was the amount of the bill?

6. Mr. A. W. Stoughton bought of Sweet & Co., July 30, 1877, the following items: 12 Arithmetics @ 60 *cts.*, 20 Geographies @ 72 *cts.*, 37 Grammars @ 43 *cts.*, 4 reams paper @ $2.25, and 3 boxes pens @ $1.10; what was the amount of the bill?

BALANCING ACCOUNTS.

104. An **Account** is a written statement of items of *debt* and *credit.* A book in which these items are recorded in separate columns, is called a **ledger.**

The ordinary form of a ledger account is shown below:

Dr.   JOHN IRVING *in acct. with* HENRY HOLT.   *Cr.*

1877			$	c.	1897			$	c.
Apr. 4	To 14 *yds.* silk	@ $1.75	24	50	May 1	By 25 *bu.* corn	@ $1.10	27	50
May 7	" 16 *yds.* cloth	" 3.50	56	00	May 19	" 46 *lbs.* butter	" .25	11	50
July 1	" 42 *yds.* muslin "	.16	6	72	July 25	Bal. due		49	34
July 7	" 8 pieces tape "	.14	1	12					
	Amount		88	34				88	34

NOTE.—The items for which **Henry** Holt is indebted to John Irving, who makes the account, are placed in the column headed *Dr.*, and the items for which John Irving is indebted to Henry Holt are placed in the column headed *Cr.*; the former are called *debits*, and the latter *credits.* The balance is the amount which, entered in the proper column, will make the sums of the two columns equal.

To find the balance of a ledger account: **Find the sum of the debits and of the credits separately, and take their difference.**

EXAMPLES.

Find the balance in the following accounts:

1. Debits, $4.19, 5.25, 7.44, 17.11, 47.42, 110.76, 308.12, 114.04, 3.79, and 10.94; Credits, $3.43, 2.11, 5.27, 15.60, 108.29, 84.12, 216.58, 94.26, and 13.12.

2. Debits, $12.56, 14.92, 27.14, 110.94, 94.11, 83.88, 7.46, 4.39, 71.18, 14.43, and 1.94; Credits, $14.87, 110.72, 3.69, 11.18, 7.42, 19.78, 9.94, 12.12, and 5.55.

REVIEW QUESTIONS.

(**83.**) What is a decimal fraction? (**84.**) What is a decimal? Decimal point? (**85.**) Repeat the numeration table for decimals. How are decimal places counted? Rule for writing a decimal? (**86.**) Rule for reading a decimal? (**87.**) What are the units of U. S. currency? Their relation? How is U. S. money written? How read? (**88.**) State the fundamental principles used in treating fractions. (**89.**) Rule for reducing common fractions to decimals. (**90.**) What is the practical rule for finding approximate values of a fraction? (**92.**) Rule for addition of decimals? (**94.**) Rule for subtraction of decimals? (**96.**) Rule for multiplication of decimals? (**98.**) Rule for division of decimals? (**99.**) What is an aliquot part of a number? Name some of the aliquot parts of a dollar. (**100.**) How do you multiply by an aliquot part of 100? (**101.**) How do you find the cost of things sold by the hundred or thousand? (**102.**) Of things sold by the ton? (**103.**) What is a bill? How find its footing? (**104.**) What is an account? A debtor? A creditor? Debits? Credits? How do you find the balance of a ledger account?

COMPOUND NUMBERS

I. DEFINITIONS AND TABLES.

DEFINITIONS.

105. A Denominate Number is one whose unit is named; as, 3 *feet*, 5 *pounds*, 16 *pennyweights* (Art. **5**).

Numbers that have the same unit are of the *same denomination;* those that have different units are of *different denominations.* Thus, 3 *feet*, and 7 *feet*, are of the same denomination ; 3 *feet*, and 7 *yards*, are of different denominations.

106. A Compound Number is a denominate number whose units are of the *same kind* but of *different denominations;* as, 3 *pounds* 6 *ounces.*

Denominate numbers are of the *same kind*, when they can be expressed in terms of a common unit. Thus, 3 *pounds*, and 6 *ounces*, are of the same kind because both can be expressed in *ounces.*

If two denominate numbers are of the same kind, that which has the *greater unit* is said to be of the *higher denomination.* Thus, 3 *pounds* is of a higher denomination than 6 *ounces.*

SCALES OF COMPOUND NUMBERS.

107. The **Scale** of a compound number is a succession of numbers showing how many times the unit of each denomination is contained in the unit of the next higher denomination. Thus, in English currency, 4 *farthings* make 1 *penny*, 12 *pence* make 1 *shilling*, and 20 *shillings* make 1 *pound ;* hence, the *scale* of English currency is

4,   12,   20.

In this scale, 4 *connects* farthings and pence, 12 *connects* pence and shillings, and 20 *connects* shillings and pounds.

In United States currency the scale is

10,   10,   10,   10.

The scale of English currency is *varying ;* that of the United States is *uniform.* The scale of United States currency is called the *scale of tens,* or the *decimal scale.*

The scales of the most important compound numbers are indicated in the following tables

TABLES OF CURRENCY.

I°. UNITED STATES CURRENCY.

108. The United States currency was established by act of Congress in 1792 ; its primary unit is 1 *dollar.*

TABLE.

10 mills (*m.*) make 1 cent..........*ct.*
10 cents   "   1 dime.........*d.*
10 dimes   "   1 dollar........$.
10 dollars   "   1 eagle.........*E.*

In business transactions the terms *eagle* and *dime* are nearly obsolete ; the term *mill* is seldom used except in official reports and *in laying taxes.*

NOTE.—The currency of the Dominion of Canada is *decimal,* and like that of the United States, it is reckoned in *dollars* and *cents.*

2°. ENGLISH CURRENCY.

109. This is the national currency of Great Britain.

TABLE.

4 farthings (*far.* or *qr.*) make 1 penny......*d.*
12 pence " 1 shilling.....*s.*
20 shillings " 1 pound .... £.
21 shillings " 1 guinea......*G.*

The primary unit of this currency is 1 *pound sterling*, and the corresponding coin is called a *sovereign*. The value of the sovereign is $4.8665.

NOTE.—The sign £, like the sign $, is written before the number to which it refers; thus, £25.

3°. FRENCH CURRENCY.

110. This is the national currency of France.

TABLE.

10 centimes (*cent.*) make 1 decime.......*d.*
10 decimes " 1 franc...... *fr.*

The primary unit of this currency is 1 *franc;* its value is 19.3 *cents*, that is, $1 is equal to about 5.18 francs. French money is usually expressed in francs and decimals of a franc.

TABLES OF WEIGHT.

4°. TROY WEIGHT.

111. This is used in weighing *gold, silver,* and some kinds of precious stones.

TABLE.

24 grains (*gr.*) make 1 pennyweight...*dwt.*
20 pennyweights " 1 ounce.... ....*oz.*
12 ounces " 1 pound.........*lb. Tr.*

EXPLANATION.—The *pound Troy* is the *primary* unit of weight in Great Britain and the United States. It was declared by an act of Parliament, which took effect in 1826, that *the brass weight of one pound Troy*, then in the custody of the Clerk of the House of Commons should be the *standard,* and that all other weights should be derived from it. It was enacted that the pound Troy should contain 5,760 *grains*, and that 7,000 such grains should make a *pound avoirdupois.*

NOTE.—The English system of weights and measures has been adopted by act of Congress, and is the legal system of the United States.

2°. APOTHECARIES' WEIGHT.

112. This differs from Troy weight in the mode of subdividing the ounce; it is used in weighing medicines.

TABLE.

20 grains (*gr.*) make 1 scruple.......... Ɔ

3 scruples " 1 dram............ ʒ

8 drams " 1 ounce.......... ℥

12 ounces " 1 pound.......... ℔

3°. AVOIRDUPOIS WEIGHT.

113. This weight is used in weighing the ordinary articles of trade and commerce.

TABLE.

16 ounces (*oz.*) make 1 pound.......... *lb.*

25 pounds " 1 quarter.... .. .. *qr.*

4 quarters " 1 hundredweight ... *cwt.*

20 hundredweight " 1 ton .... *T.*

The primary unit is 1 *pound*, equal to 7,000 grains Troy.

In weighing coarse articles liable to wastage, as coal at the mines, and the like, it is customary to call 112 *lbs.* a *hundredweight*, and 28 *lbs.* a *quarter.*

TABLES OF TIME.

DEFINITION AND EXPLANATION.

114. **Time** is a measured portion of duration. Its primary unit is 1 *mean*, or *average, solar day.*

EXPLANATION.—An astronomical year is the time required for the earth to revolve about the sun; but this period does not contain an exact number of days; hence, for civil purposes, an artificial year is adopted. The length of the civil year is sometimes 365 *days*, and sometimes 366 *days*, and these are so distributed that after a long period the *average* length of the civil year is equal to that of the astronomical year.

Every year divisible by 4 (except centennial years not divisible by 400) are *leap years* and contain 366 days each; all other years are *common years* and contain 365 days each.

TABLE.

60 seconds (*sec.*) make 1 minute......... *min.*

60 minutes " 1 hour........... *hr.*

24 hours " 1 day........... *da.*

7 days " 1 week.......... *wk.*

365 days " 1 common year.... *yr.*

366 days " 1 leap year....... *yr.*

100 years " 1 century........ *C.*

The year is divided into 12 parts called *months.* Their order and the number of days in each is shown in the

TABLE.

1°. January........31 days.	7°. July..........31 days.
2°. February...28 or 29 "	8°. August.......31 "
3°. March.........31 "	9°. September ....30 "
4°. April..........30 "	10°. October.......31 "
5°. May...........31 "	11°. November.....30 "
6°. June.........,..30 "	12°. December.....31 "

In common years February has 28 *days*; in leap years 29.

MEASURES OF LENGTH.

DEFINITIONS.

115. A **Magnitude** is anything that can be *measured ;* that is, anything that can be expressed in terms of a thing of the same kind taken as a *unit.*

116. A **Line** is a magnitude that has *length,* without *breadth* or *thickness.*

A **curved line** is one whose direction changes at every point ; as GH.

A **straight line** is one whose direction does not change at any point; as AB.

Straight lines are **parallel** when they have the same direction ; as CD and EF.

The **length** of a line is the number of times it contains a given straight line taken as a *unit.*

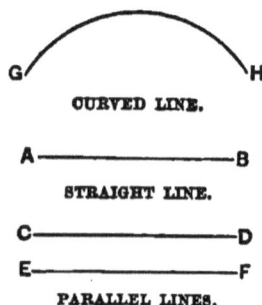

CURVED LINE.

STRAIGHT LINE.

PARALLEL LINES.

I°. LONG MEASURE.

117. This is used for measuring distances and dimensions of objects.

TABLE.

12	inches (*in.*)	make	1 foot......*ft.*
3	feet	"	1 yard......*yd.*
5½	yards	"	1 rod.......*rd.*
40	rods	"	1 furlong...*fur.*
8	furlongs, or 320 rods	"	1 mile......*mi.*
3	miles	"	1 league....*lea.*

It is found convenient in practice to reduce *yards, feet,* and *inches* to *half yards* and *inches ;* we thus avoid the inconvenience of a scale in which one of the numbers is fractional. In this case, the first part of the preceding table may be replaced by the following :

18 inches	make	1 half yard......*hf. yd.*
11 half yards	"	1 rod............*rd.*

The **yard** is the primary unit of English and American measures of length. By the act of Parliament already referred to, it was declared that the brass standard yard, then in the custody of the Clerk of the House of Commons, should be the *imperial standard yard.* From it we derive all other measures of length.

NOTE.—In measuring *cloth, ribbons,* and the like, the yard is sub- divided into *halves, quarters, eighths,* and *sixteenths.*

2°. SURVEYORS' MEASURE.

118. This is used in measuring land. The unit is a **Gunter's Chain**; this chain is 4 *rods,* or 66 *feet,* in length, and is divided into 100 equal parts, called *links.*

TABLE.

7.92 inches make 1 link........*li.*
100   links   "   1 chain.......*ch.*
80   chains   "   1 mile........*mi.*

MEASURES OF SURFACE.
DEFINITIONS.

119. A **Surface** is a magnitude that has *length* and *breadth,* without *thickness.*

A **Plane** is a surface such that if a straight line is applied to it in any direction it will coincide with the surface throughout.

120. An **Angle** is the opening between two lines that meet at a point; as, BAC.

The lines AB and AC are called *sides,* and the point A is called the *vertex* of the angle.

If one straight line meets another so as to make the two adjacent angles equal, the first line is said to be **perpendicular** to the second, and the angles are called **right angles**; thus, if the angles BAD and BAC are equal, they are both right angles and BA is perpendicular to DC.

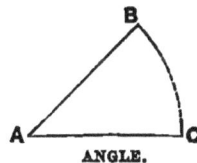
ANGLE.

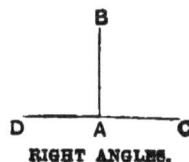
RIGHT ANGLES.

121. A **Square** is a plane figure bounded by four equal lines, called **Sides**, and having all its angles right angles; as, ABCD.

A **Rectangle** is bounded by four lines, parallel two and two, and having all its angles right angles.

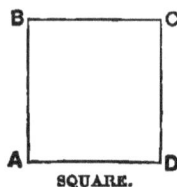

SQUARE.

122. A **Unit of Surface** is a square whose sides are equal to the *unit of length*. If each side is a yard, the square is called a *square yard*.

The **Area of a Surface** is an expression for that surface in terms of *a square unit*.

NOTE.—It is shown in geometry that the area of a rectangle is equal to the product of its *length* by its *breadth;* that is, the number of *square units* in the surface is equal to the number of units in its length multiplied by the number of units in its breadth.

I°. SQUARE MEASURE.

123. This is used in measuring surfaces.

TABLE.

144 square inches (*sq. in.*) make 1 square foot....*sq. ft.*

  9 square feet   "   1 square yard....*sq. yd.*

30¼ square yards   "   1 square rod.....*sq. rd.*

160 square rods   "   1 acre.. ....*A.*

It is found convenient in practice to reduce *square yards*, *square feet*, and *square inches*, to *quarter square yards* and *square inches*, to avoid using a scale containing a fractional number.   In this case

324 square inches   make 1 quarter square yard..*qr.sq. yd.*

121 square quarters   "   1 square rod..........*sq. rd.*

160 square rods   "   1 acre................*A.*

2°. LAND MEASURE.

124. This is used in measuring land.

TABLE.

10000 square links (*sq. li.*) make 1 square chain....*sq. ch.*
  10 square chains   "   1 acre....... ...*A.*
  640 acres   "   1 square mile.....*sq. mi.*

The area of land is also reckoned by the following

TABLE.

  40 square rods (*sq. rds.*) make 1 rood......*R.*
  4 roods   "   1 acre.......*A.*

In government surveys, a square mile is called a *section;* 36 sections make one township.

MEASURES OF VOLUME AND CAPACITY.

DEFINITIONS.

125. A **Volume** is a magnitude that has *length, breadth,* and *thickness* or *height.*

126. A **Cube** is a volume or solid, bounded by six equal squares, called **Faces** ; the sides of the squares are called **Edges** of the cube. Thus, ABCD–E is a cube, ABCD is one of its *faces,* AB, BC, and BE are *edges.*

CUBE.

A rectangular volume, or solid, whose edges are not equal is a *parallelopipedon.*

PARALLELOPIPEDON.

127. A **Unit of Volume** is a cube whose edges are equal to the unit of length. If each edge is a lineal yard, the cube is called a *cubic yard.*

128. The **Content** of a volume is an expression for that volume in terms of a cubic unit.

NOTE.—It is shown in geometry that the content of a cube or of any rectangular solid is equal to the product of its *three dimensions;* that is, the number of *cubic units* in the volume is equal to the number of units in its *length,* multiplied by the number of units in its *breadth,* multiplied by the number of units in its *height.*

129. A **Unit of Capacity** is a measure having a determinate content, or capacity.

1°. CUBIC MEASURE.

130. This is used in measuring volumes or solids.

TABLE.

1728 cubic inches (*cu. in.*) make 1 cubic foot.....*cu. ft.*.

27 cubic feet   "   1 cubic yard.....*cu. yd.*

A Cord of Wood is a pile 4 *ft.* wide, 4 *ft.* high, and 8 *ft.* long. A foot in length from such a pile is called a Cord Foot. A cord foot is 16 cubic feet.

TABLE.

16 cubic feet (*cu. ft.*)   make 1 cord foot.... *C. ft.*

8 cord feet or 128 *cu. ft.*   "   1 cord........ *C.*

2°. DRY MEASURE.

131. This is used in measuring dry articles, as grain, fruit, salt, and the like.

TABLE

2 pints (*pt.*) make 1 quart........*qt.*

8 quarts   "   1 peck.........*pk.*

4 pecks   "   1 bushel.......*bu.*

The primary unit is 1 *bushel*. The bushel (known as the Winchester bushel) is a cylindrical measure 18½ inches across and 8 inches deep ; it contains 2,150⅖ cubic inches *nearly*.

3°. LIQUID MEASURE.

132. This is used in measuring liquids.

TABLE.

4	gills (*gi.*)	make	1 pint........*pt.*
2	pints	"	1 quart........*qt.*
4	quarts	"	1 gallon......*gal.*
31½	gallons	"	1 barrel.......*bbl.*
2	barrels	"	1 hogshead...*hhd.*
2	hogsheads	"	1 pipe........*pi.*
2	pipes	"	1 tun........*tun.*

The primary unit is 1 *gallon;* it contains 231 *cubic inches*. The *liquid quart* is about ⅚ of a *quart of dry measure*.

ANGULAR MEASURE AND LONGITUDE.

DEFINITIONS.

133. A **Circle** is a plane figure bounded by a curved line every point of which is equally distant from a point within called the **Centre**. The bounding line is called the **Circumference**, and any part of this circumference is called an **Arc** of the circle.

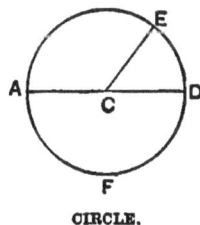

134. A **Diameter** is a line passing through the centre and terminating in the circumference.

135. A **Radius** is a line drawn from the centre to any point of the circumference.

Thus, AEDF is a circle, C its centre, AD a diameter, and CE a radius.

I°. ANGULAR MEASURE.

136. In measuring angles, the **Right Angle** (Art. **120**) is taken as the *primary unit*. The ninetieth part of a right angle is called a **Degree**.

TABLE.

60 seconds (″) make 1 minute........
60 minutes   "   1 degree .... °
90 degrees   "   1 right angle .... *rt. a.*

DEFINITION AND EXPLANATION.

137. The **Longitude** of a place is the angular distance of the meridian of that place from some *standard meridian*. It is measured on the equator, and is equal to the angle through which the earth turns on its axis whilst the sun is passing from the meridian of one place to that of the other. But the earth turns through an angle of 360° in 24 *hours*, that is, it turns through an angle of 15° in 1 *hour;* hence, we have the following relations between the difference of longitude of two places and the difference of their local times.

TABLE.

15° of arc make 1 *hour* of time.
15′ of arc   "   1 *minute* of time.
15″ of arc   "   1 *second* of time.

NOTE.—All longitudes referred to in this book are reckoned from the meridian of Greenwich, England.

MISCELLANEOUS TABLES.

138. The following miscellaneous tables are often used in operating on compound numbers:

1°. COUNTING.

12 things make 1 dozen............ *doz.*

12 dozen   "   1 gross............. *gr.*

12 gross   "   1 great gross .... *g. gr.*

20 things   "   1 score............ *sc.*

2°. PAPER.

24 sheets make 1 quire............. *qr.*

20 quires   "   1 ream............. *ream.*

2 reams   "   1 bundle........... *bund.*

2 bundles "   1 bale............. *bale.*

3°. BOOKS.

A book in which
each sheet is folded into 2 leaves is a folio.

"   "   "   "   4   "   "   quarto, or 4to.

"   "   "   "   8   "   "   octavo, or 8vo.

"   "   "   "   12   "   "   duodecimo, or 12mo.

"   "   "   "   16   "   "   16mo.

"   "   "   "   24   "   "   24mo.

"   "   "   "   32   "   "   32mo.

THE METRIC SYSTEM.

DEFINITIONS AND EXPLANATIONS.

139. The **Metric System** is a system of weights
and measures based on a primary unit of length called a
Meter.

This system was first adopted by France and afterwards by various other countries. Since 1866 its use has been permitted in the United States, by act of Congress.

The *meter* is approximately equal to $\frac{1}{10000000}$ of the
distance from the equator to the north pole, measured on

the meridian through Paris. It is nearly equal to 39.37 inches.

The *scales* of all compound numbers in the metric system are *decimal*.

The names of units in the ascending scale are formed by prefixing the following numerals to the names of the primary units: *deca*, ten ; *hecto*, one hundred ; *kilo*, one thousand ; and *myria*, ten thousand.

The names of units in the descending scale are formed by prefixing the following numerals to the names of the primary units: *deci*, one tenth ; *centi*, one hundredth ; and *milli*, one thousandth.

MEASURES OF LENGTH.

140. The primary unit is the **meter.**

TABLE.

10 millimeters (*mm.*) make	1 centimeter	(*cm.*)	=	.3937 *in.*
10 centimeters "	1 decimeter	(*dm.*)	=	3.937 "
10 decimeters "	1 meter	(*m.*)	=	39.37 "
10 meters "	1 decameter	(*decam.*)	=	393.7 "
10 decameters "	1 hectometer	(*hectom.*)	=	328 *ft.* 1 *in.*
10 hectometers "	1 kilometer	(*kilom.*)	= 3280 *ft.* 10 *in.*

Lengths are usually expressed *decimally* in terms of some one of the units of the preceding scale. Small distances are expressed in *millimeters*, ordinary distances in *meters*, and long distances in *kilometers*.

MEASURES OF SURFACE.

141. The primary unit of ordinary surfaces is a Square Meter.

TABLE.

100 sq. centimeters make 1 sq. decimeter = 15.5 *sq.in.*
100 sq. decimeters " 1 sq. meter = 1550 " "

For *land measure* the primary unit is the *are* (pron. *ar*) ; it is a square decameter, or 100 square meters.

100 sq. meters make 1 are $\quad = 119.6 \quad$ *sq. yds.*
100 ares $\qquad$ " $\quad$ 1 hectare $= \quad$ 2.471 *acres.*

MEASURES OF VOLUME AND CAPACITY.

142. The primary unit of ordinary volumes is the Stere (pron. *stair*); it is a **Cubic Meter.**

TABLE.

1000 cu. centimeters make 1 cu. decimeter $= 61.026$ *cu. in.*
1000 cu. decimeters $\quad$ " $\quad$ 1 cu. meter $\quad = 35.316$ *cu. ft.*

For *wood measure* the primary unit is the *stere.*

10 decisteres make 1 stere $(st.) = .2759$ *cords.*
10 steres $\qquad$ " $\quad$ 1 decastere $= 2.759$ $\quad$ "

For measures of capacity the primary unit is the liter (pron. *leeter*). It is a cubic *decimeter.*

TABLE.

				Dry Measure.	Liquid Measure.
10 centiliters (*cl.*)	make	1 deciliter	(*dl.*)	$= .0908$ *qt.* $=$	.1057 *qt.*
10 deciliters	"	1 liter	(*l.*)	$= .908$ *qt.* $=$	1.0567 *qt.*
10 liters	"	1 decaliter	(*decal.*)	$= 9.08$ *qts.* $=$	2.6417 *gals.*
10 decaliters \	"	1 hectoliter	(*hectol.*)	$= 2.8375$ *bu.* $=$	26.417 *gals.*

MEASURES OF WEIGHT.

143. The primary unit of weight is the **Gram.** It is the weight of a *cubic centimeter* of distilled water at 39° Fah.

TABLE.

10 milligrams (*mg.*)	make	1 centigram	(*cg.*)	$=$	.1543 *grs.*
10 centigrams	"	1 decigram	(*dg.*)	$=$	1.5432 "
10 decigrams	"	1 gram	(*g.*)	$=$	15.432 "
10 grams	"	1 decagram	(*decag.*)	$=$	.3527 *oz. av.*
10 decagrams	"	1 hectogram	(*hectog.*)	$=$	3.5274 "
10 hectograms	"	1 kilogram	(*kilog.*)	$=$	2.2046 *lbs. av.*

Small weights are expressed in *milligrams* and large ones in *kilograms*. The kilogram is the weight of a liter of water at 39° Fah.

A *quintal* is 100 kilograms or 220.46 *lbs. av.* The French *tonneau* (pron. *ton-no*), or metric ton, is the weight of a cubic meter of water, or to 2,204 *lbs. av.*

We may read a number in the metric system in terms of all its units, or in terms of any one of them. Thus, the expression 27.34 *meters* may read 2 *decameters*, 7 *meters*, 3 *decimeters*, and 4 *centimeters*, but it is usually read as it is written, 27 *and* 34 *hundredths of a meter*.

REVIEW QUESTIONS.

(105.) What is a denominate number? When are numbers of the same denomination? Of different denominations? **(106.)** What is a compound number? When are numbers of the same kind? Of different kinds? **(107.)** What is a scale? Give the scale of English currency? Of United States currency? **(111.)** What is the primary unit of weight for Great Britain and the United States? How was it established? How many grains in a pound Troy? How many in a pound avoirdupois? **(114.)** What is time? Explanation of the lengths of years and of their distribution? What years are leap years? What years are common years? Name the months and the number of days in each. **(115.)** Define magnitude. **(116.)** Define a line. What is the length of a line? **(119.)** Define a surface. **(120.)** Define an angle. A right angle. **(121.)** Define a square. **(122.)** Define a unit of surface. Area. **(125.)** Define a volume. **(126.)** Define a *circle*. **(127.)** Define a unit of volume. Content of a volume. Unit of capacity. **(130.)** What is a cord of wood? What is a cord foot? **(133.)** Define a circle. Circumference. Arc. **(134.)** Define a diameter. **(135.)** Define a radius. **(136.)** What is a degree? **(137.)** Define longitude. What is the relation between longitude and time, and how was it established? **(139.)** What is the metric system? A meter? How are the names of metric numbers formed? **(141.)** What is an *are*? For what purpose is it used? **(142.)** What a *stere*? A *liter*? **(143.)** What is a gram?

II. REDUCTION.

DEFINITIONS.

144. **Reduction of denominate numbers** is the operation of changing a number from one denomination to another, without altering its value.

A number may be changed from a *higher* to a *lower* denomination, or from a *lower* to a *higher* denomination; in the former case the operation is called reduction descending, in the latter it is called reduction ascending.

REDUCTION DESCENDING.

145. **Reduction descending** is the operation of changing a number from a higher to a lower denomination.

MENTAL EXERCISES.

1. How many *pence* are there in 1 *shilling?* How many in 5s.? in 14s.? In 1 *foot*, how many inches? in 9 *ft.?* in 16 *ft.?* In 1 *hour*, how many *minutes?* in 3 *hrs.?* in 9 *hrs.?* in 15 *hrs.?* in 18 *hrs.?*

2. How many *ounces* in 1 *pound* avoirdupois? in 5 *lbs.?* in 8 *lbs.?* in 9 *lbs.?* In 1 *pound sterling*, how many *shillings?* in £6? in £11? in £14? In 5 *yards*, how many *inches?* in 12 *yds.?* in 14 *yds.* 8 *in.?*

EXPLANATION.—In 14 *yds.* there are 14 × 12, or 168 *inches;* hence, in 14 *yds.* 5 *in.* there are 168 + 5, or 173 *inches.*

3. In 7 *shillings*, how many *pence?* in 7s. 6d.? in 9s. 9d.? In 3 *pecks*, how many *quarts?* in 2 *pks.* 7 *qts.?* in 3 *pks.* 5 *qts.?* In 4 *hundredweight*, how many *quarters?* in 9 *cwt.* 3 *qrs.?* in 12 *cwt.* 2 *qrs.?* In 7 *meters*, how many *decimeters?* in 8 *m.* 7 *dm.?* in 8.7 *m.?* in 11.4 *m.?* In 8 *dollars*, how many *cents?* in $9.50? in $14.15?

4. In 4 *yds.*, how many *feet?* How many *inches?* In 6 *yds.*, how many *inches?* in 5 yds. 4 *in.?* in 7 yds. 9 *in.?* How many *inches* in 6 yds. 2 *ft.* 3 *in.?*

EXPLANATION.—In 6 *yds.* 2 *ft.* there are 20 *ft.*, and in 20 *ft.* 9 *in.* there are 249 *inches.*

5. In £2 10s., how many *shillings?* how many *pence?* In £3 4s., how many *pence?* How many pints in 7 *qts.?* in 6 *qts.* 1 *pt.?* in 2 *pks.* 2 *qts.?* in 3 *pks.* 4 *qts.* 1 *pt.?* In 4 *meters*, how many *decimeters?* how many *centimeters?* how many *millimeters?* In $4.15, how many *mills?* in $5.07? in $4.723?

OPERATION OF REDUCTION DESCENDING.

146. Let it be required to reduce £7 4s. 3d. to pence:

EXPLANATION.—The given number contains 7 units, each equal to £1, and from the scale (Art. **109.**) we see that each pound contains 20s.; multiplying 7 by 20 and adding 4, we have 144, that is, £7 4s. is equal to 144s.; but from the scale we see that each shilling is equal to 12d.; mul-

OPERATION.

£7   4s.   3d.

20

‾‾‾‾

144s

12

‾‾‾‾

1731d. · *Ans.*

tiplying 144 by 12 and adding 3 we have 1731; that is, the given number is equal to 1731d.

Since all similar cases may be treated in the same manner, we have the following

RULE.

I. Multiply the units of the highest denomination by the number of the scale that connects this denomination with the one next lower and add the units of the latter denomination to the product.

II. Multiply this result by the number that

connects it with the next lower denomination and add the units of that denomination to the product.

III. Continue this operation till the required denomination is reached.

EXAMPLES.

1. How many farthings are there in £4 11*s.* 3*d.* 2*far.?*
2. How many grains in 3 *lbs.* 8 *oz.* 6 *dwts.* 4 *grs.?*
3. How many minutes in 3 *wks.* 5 *da.* 5 *hrs.?*
4. How many yards in 3 *mi.* 28 *rds.* 2 *yds.* ?
5. How many square feet in 40 square rods ?
6. How many pints in 8 *bu.* 3 *pks.* 2 *qts.* 1 *pt.* ?
7. How many seconds in 4 *wks.* 3 *da.* 17 *hrs.* ?
8. In £31 8*s.* 9¼*d.*, how many farthings ?
9. How many inches in 6 *rds.* 4 *yds.* 2 *ft.* 9 *in.* ?
10. In 17.5 square chains, how many square links ?
11. In 25 cords of wood, how many cord feet and how many cubic feet ?   In 172 *C.* ?   In 115 *C.* 8 *C. ft.* ?
12. How many quarts are there in 20 *bu.* 3 *pks.?*
13. Reduce £⅔ to shillings and pence.

EXPLANATION.—Multiplying by 20 to reduce *pounds* to *shillings,* we have £⅔ = ¹⁰⁰⁄₆*s.* = 16⅔*s.*; multiplying ⅔*s.* by 12 to reduce *shillings* to *pence,* we have ⅔*s.* = ²⁴⁄₃*d.* = 8*d.*: hence, £⅔ = 16*s.* 8*d. Ans.*

In like manner we may reduce any denominate fraction to integral units of a lower denomination ; hence the following

RULE.

Multiply the fraction by the number that connects it with the next lower denomination ; then multiply the fractional part of the result by the number that connects it with the next lower denomination, and so on.

14. Reduce .714 *yds.* to feet and inches.

SOLUTION.—Multiplying by 3, we have .714 *yds.* = 2.142 *ft.*; multiplying .142 *ft.* by 12, we have .142 *ft.* = 1.704 *in.*; hence, .714 *yds.* = 2 *ft.* 1.704 *in.* *Ans.*

Reduce the following *denominate fractions* to integral units of different denominations:

15. £.875.	24. 2⅞¼ *mi.*	33. £⅔⅓.
16. .9375 *lbs. Tr.*	25. 7¾⁴⁷ *fur.*	34. 4⁷₁₅ *bu.*
17. .11875 *tons.*	26. 3½⅔ *bu.*	35. 2.56 *T.*
18. .45 *sq. yds.*	27. 43³₁₆ *yds.*	36. 3.567 *da.*
19. £2¾.	28. 17¾ *yds.*	37. .574 *mi.*
20. 3⁴⁴ *da.*	29. 4.562 *hhds.*	38. $.895.
21. 141¾ *wks.*	30. 3½⁴°.	39. 3.46 *A.*
22. 4¹¹₂₀ *T.*	31. 12°.162.	40. .457 *gals.*
23. 3¹²³³³₁₆₀₀₀ *T.*	32. .436 *T.*	41. ⁷⁹⁷₁₇₆₀ *mi.*

REDUCTION ASCENDING.

147. Reduction Ascending is the operation of changing a number from a lower to a higher denomination.

MENTAL EXERCISES.

1. In 12 *inches* how many *feet?* In 96 *in.* how many *ft.?* in 144 *in.?* in 204 *in.?* In 120 *seconds* how many *minutes?* in 960 *sec.?* In 16 *qts.* how many *pks.?* in 96 *qts.?* How many *minutes* in 240″? in 540″? How many *degrees* in 300′? in 540′? Reduce 600′ to *degrees.*

2. In 64 *oz. av.,* how many *pounds?* In 100 *oz. av.,* how many *pounds* and how many *ounces* remain? Reduce 110 *oz. av.* to *pounds.* *Ans.* 6 *lbs.* 14 *oz.* In 219 *in.* how many *feet* and *inches?* Reduce 154 *in.* to a higher denomination. *Ans.* 12 *ft.* 10 *in.*

3. In 64 *pks.* how many *bu.?* In 74 *pks.* how many *bu.* and how many *pks.* remain? In 69 *pks.* how many *bu.* and *pks.?* Reduce 127 *qts.* to *pks.* Reduce 117 *d.* to *shillings* and *pence.* Reduce 121 *inches* to *feet* and *inches.*

4. In 145 *d.* how many *shillings* and how many *pence?* In 163 *inches* how many *feet?* *Ans.* 13 *ft.* 7 *in.* In 13 *ft.* how many *yds.?* *Ans.* 4 *yds.* 1 *ft.* How many *yds. ft.* and *in.* in 163 *in.?* *Ans.* 4 *yds.* 1 *ft.* 7 *in.* Reduce 213 *in.* to higher denominations. Express 255 *pts.* in higher denominations. Reduce 280 *d.* to higher denominations.

OPERATION OF REDUCTION ASCENDING.

148. Let it be required to reduce 1,731 *pence* to *pounds, shillings,* and *pence:*

EXPLANATION.—We see from the scale, (Art. **109**), that 12 connects pence with shillings; dividing 1,731 by 12, we find a quotient 144 with a remainder 3; hence the given number is equal to 144 *s.* 3 *d.*; in like manner dividing 144 by 20, we find a quotient

OPERATION.

$$12)1731d.$$
$$\overline{20)144s....3d.}$$
$$\overline{\pounds7 \ ...4s.}$$
$$\pounds7 \ 4s. \ 3d. \quad Ans.$$

7 and a remainder 4, that is, 144 *s.* is equal to £7 4 *s.*; hence, the given number is equal to £7 4 *s.* 3 *d.*

In like manner we may treat all similar cases; hence, the following

RULE.

I. Divide the units of the given denomination by the number of the scale that connects this denomination with the one next higher; the remainder will be units of the same denomination as the dividend.

II. Divide the quotient by the number that connects it with the next higher denomination; the

6

remainder will be units of the same denomination as the new dividend.

III. Continue the operation till the required denomination is reached.

EXAMPLES.

1. Reduce 15,732 grains to pounds Troy.
2. Reduce 525,960 minutes to weeks.

PROOF.—Reduction descending is proved by reduction ascending, and the reverse; thus, the preceding example is proved by reducing 52 *wks.* 1 *da.* 6 *hrs.* to minutes; this gives 525,960 *min.*

3. Express 49,180 *grains* in *pounds*, Apothecaries' weight.
4. Reduce 3,392 *grains* to *ounces*, Apothecaries' weight.
5. Reduce 2,945 *inches* to *higher denominations.*

EXPLANATION.—In this case we reduce 2,945 inches to *half yards;* then we reduce the *half yards* to rods, (Art. **117**). Dividing 2,945 *inches* by 18, because there are 18 *inches* in a half yard, we find 163 *half yards* and a remainder of 11 *inches;* dividing 163 by 11, because there are 11 *half yards* in a rod, we find 14 *rods* with a remainder equal to 9 *half yards,* or to 4 *yards* and 18 *inches;* adding the 18 *inches* to the first remainder, 11 *inches,* we find 29 *inches,* or 2 *ft.* 5 *in.* Hence, 2,945 *in.* = 14 *rds.* 4 *yds.* 2 *ft.* 5 *in.*

Reduce

6. 1,365 *in.* to *rods.*
7. 272,668 *in.* to *miles.*
8. 88,435 *in.* to *miles.*
9. 873 *oz. av.* to *quarters.*
10. 7,634 *gi.* to *gallons.*
11. 8,372 *far.* to *pounds.*
12. 14,311 *gr.* to *lbs. Tr.*
13. 4,771 ℈ to *lbs.*

Reduce

14. 16,411 *oz.* to *cwts.*
15. 311,375 *sec.* to *days.*
16. 31,463 *min.* to *weeks.*
17. 21,118″ to *degrees.*
18. 1,114′ to *degrees.*
19. 4,643 *mm.* to *meters.*
20. 13,362 *sq. in.* to *sq. yds.*
21. 1,211,312 *sq. li.* to *acres.*

22. Reduce 7 *s.* 6 *d.* to a fractional part of £1.

EXPLANATION.—Reducing 7s. 6d. to pence, we have 7s. 6d. = 90d.; reducing the given unit £1 to pence, we have, £1 = 240d.: hence, 7s. 6d. = £$\frac{90}{240}$ = £$\frac{3}{8}$. *Ans.* In like manner all compound numbers may be expressed in fractional parts of a higher integral unit; hence the following

RULE.

Reduce the compound number to its lowest denomination; also reduce the given unit to the same denomination; then divide the former by the latter.

23. What part of a tun is 3 *hhds.* 31 *gals.* 2 *qts.?*

24. What part of a hogshead is 3 *gals.* 2 *qts.?*

25. What part of a mile is 116 *rds.* 2 *yds.?*

26. What part of a right angle is 3° 15′ 12″ ?

27. What part of a gallon is 2 *qts.* 1 *pt.* 1 *gi.?*

28. What part of a cord is 4 *C. ft.* 7 *cu. ft.?*

29. What part of a *meter* is 714 *mm.?*

30. What *decimal of a ton* is 15 *cwt.* 3 *qrs.* 2½ *lbs.?*

EXPLANATION.—For convenience the units of the several denominations are written in a vertical column, the least denomination at the top, and the numbers of the corresponding scale are written on the left from the top downward. Dividing 2½ *lbs.*, or its equal 2.5 *lbs.*, by 25,

OPERATION.

25 | 2.5 *lbs.*

4 | 3.1 *qrs.*

20 | 15.775 *cwt.*

.78875 *T.*

to reduce it to quarters, we have .1 *qr.*, which being annexed to 3 *qrs.* gives 3.1 *qrs.*; dividing 3.1 *qrs.* by 4, to reduce it to hundreds, we have .775 *cwt.*, which being annexed to 15 *cwt.* gives 15.775 *cwt.*; dividing this by 20 to reduce it to tons, we have .78875 *T.*, which is the answer.

In like manner any compound number may be reduced to a decimal of a higher denomination; hence the following

RULE.

Divide the units of the lowest denomination by the number of the scale that connects this · de-

nomination with the next higher one, (Art. **107***), and annex the quotient to the units of that denomination; then divide this result by the number that connects it with the next higher denomination and annex the quotient to the units of that denomination; and so on, till the required denomination is reached.*

31. In 5 *bu.* 3 *pks.* 6 *qts.*, how many *pecks?*

EXPLANATION.—We first reduce 5 *bu.* 3 *pks.* to *pks.*, which gives 23 *pks.;* we then reduce 6 *qts.* to a decimal of a *pk.*, which gives .75 *pks.;* hence, 5 *bu.* 3 *pks.* 6 *qts.* = 23.75 *pks.*, *Ans.*

Reduce the following to the units indicated:

32. £14 17*s.* 3*d.* to *pounds.*

33. 7 *mi.* 281 *rds.* to *miles.*

34. 2*cwt.* 1*qr.* 12½*lbs.* to *tons.*

35. 3¼ ℥ to *lbs.* (*Apoth.*)

36. 2⅓ *pts.* to *gallons.*

37. 21 *in.* to *yds.*

38. 2 *lbs.* 4 *oz.* 5 *dwts.* to *oz.*

39. 2 *T.* 4 *cwt.* 3 *qrs.* to *lbs.*

40. 8 *bu.* 3 *pks.* 6 *qts.* to *pks.*

41. £5 4*s.* 9*d.* to *shillings.*

42. 3 *bbls.* 14⅞ *gals.* to *gals.*

43. 47 *rds.* 4⅜ *yds.* to *yds.*

44. 47° 35′ 42″ to *minutes.*

45. 4*m.* 3*dm.* to *decameters.*

46. What part of £3 10*s.* 6*d.* is £1 15*s.* 4*d.* ?

EXPLANATION.—Here we reduce both numbers to *pence*, the lowest denomination named in either ; we then perform the operation indicated in the question.   Thus, £3 10*s.* 6*d,* = 846*d.*, and £1 15*s.* 4*d.* = 424*d.;* hence the required answer is ⁴²⁴⁄₈₄₆ or ²¹²⁄₄₂₃.

47. What part of 36° 15′ 20″ is 11° 13′ 20″ ?

48. What part of 4 *gals.* 3 *qts.* is 2 *gals.* 3 *qts.* ?

49. What part of 2 *T.* 6 *cwt.* 20 *lbs.* is 7 *cwt.* 10 *lbs.* ?

50. What part of £⅜ is 1½*s.* ?

51. What part of ⅞ *mi.* is 4¾ *rds.* ?

52. What part of 3 *bu.* 3 *pks.* is 2 *bu.* 1 *pk.* ?

MISCELLANEOUS EXAMPLES IN REDUCTION.

1. How many grains in 1 *lb.* 11 *oz.* 15 *dwts.?*
2. In 97,397 grains of gold how many pounds Troy?
3. In 24 *tons* 17 *cwt.* 3 *qr.* how many pounds?
4. In 136½ bushels how many quarts?
5. How many miles in 1,571,328 inches?
6. In 2,624 cubic feet of wood, how many cords?
7. Reduce 18,545,435 *sec.* to days. To weeks.
8. What is the value of ⅝ of 1*s.?*   Of ⅓ of 6*s.?*

NOTE.—Metric weights and measures may be converted into English weights and measures by the tables of equivalents given in Arts. **140–143.**

9. How many inches in 7 meters?   In 13 meters?
10. How many meters in 605 inches?   In 319 *in.?*
11. How many miles in 12 kilometers?
12. How many kilometers in 15 miles?   In 7½ *mi.?*
13. How many quarts (liquid measure) in 41 liters?
14. How many gallons in 4 hectoliters?
15. How many liters in 31½ gallons?   In 74 *galls.?*
16. How many ounces in 711 grammes?   In 51 *g.?*
17. How many pounds in 74 kilogrammes?
18. Reduce 510 kilogrammes to pounds.
19. How many cords in 15 steres of wood?
20. Reduce 17½ steres of wood to cords.   To *cu. ft.*
21. Reduce 17*s.* 9¾*d.* to a decimal of a pound.
22. What part of £1 2*s.* 6*d.* to 1*s.* 11*d.?*
23. Reduce 1 *lb.* 9 *oz.* 15 *gr.* to pounds.   To ounces.
24. Reduce 3 *da.* 14 *hrs.* 25 *min.* to weeks.
25. What part of a barrel is 8 *gals.* 2 *qts?*
26. What part of a mile is 74 *rds.* 5 *yds.?*

27. What part of a cord is 3¾ cord feet?

28. What part of a pound sterling is 5s. 9d.?

29. Reduce £.187 to lower denominations.

30. Reduce .574 miles to lower denominations.

REVIEW QUESTIONS.

(144.) What is reduction of denominate numbers? How many kinds of reduction? **(145.)** What is reduction descending? **(146.)** Rule for reduction descending? Rule for reducing denominate fractions to integral units? **(147.)** What is reduction ascending? **(148.)** Rule for reduction ascending? Rule for reducing compound numbers to any unit of the kind named? Rule for finding the part that one compound number is of a similar number?

III. ADDITION OF COMPOUND NUMBERS.

DEFINITION.

149. Addition of Compound Numbers is the operation of finding the **sum** of two or more compound numbers of the same kind.

MENTAL EXERCISES.

1. What is the sum of 9 *inches* and 11 *inches?* Of 13 *inches* and 22 *inches?* Of 8 *ft.* and 17 *ft.?* What is the sum of 18d. and 16d.? Of 10d., 7d., and 19d.?

2. What is the sum of 7 *ft.* and 8 *ft.?* How many yards in 7 *ft.?* In 8 *ft.?* In the sum of 7 *ft.* and 8 *ft.?* What then is the sum of 2 *yds.* 1 *ft.* and 2 *yds.* 2 *ft.?* What is the sum of 11 *qts.* and 15 *qts.?* How many gallons in 11 *qts.?* In 15 *qts.?* In 11 *qts.* + 15 *qts.?* What then is the sum of 2 *gals.* 3 *qts.*, and 3 *gals.* 3 *qts.*

3. What is the sum of 3 *gals.* and 5 *gals.*? Of 2 *qts.* and 3 *qts.*? Of 3 *gals.* 2 *qts.*, and 5 *gals.* 3 *qts.*? What is the sum of 3s. 8d., and 5s. 9d.? Of 3 *yds.* 2 *ft.*, 5 *yds.* 1 *ft.*, and 7 *yds.* 2 *ft.*?

NOTE.—The principles used in the *addition of compound numbers* are the same as those used in the *addition of simple numbers.*

OPERATION OF ADDITION OF COMPOUND NUMBERS.

150. Let it be required to find the sum of £7 4s. 3d., £11 9s. 8d., and £14 12s. 9d.:

Explanation. —We write the numbers so that units of the same denomination shall stand in the same column. The sum of the numbers in the first column is 20d., or 1s. 8d. (Art. **147**); setting down 8d., we carry forward 1s., and add it to the second column.

OPERATION.

£	s.	d.
7	4	3
11	9	8
14	12	9
£33	6s.	8d.

The sum of the numbers in the second column, thus increased, is 26s., or £1. 6s.; setting down 6s., we carry forward £1 to the next column, which then amounts to £33. The required sum is therefore £33 6s. 8d.

In like manner we may treat all similar cases ; hence, the following

RULE.

I. Write the numbers so that units of the same denomination shall stand in the same column.

II. Add the units of the lowest denomination and divide their sum by the number of the scale that connects this denomination with the next higher one; set down the remainder and carry the quotient to the next column.

III. Add the units of the second column thus increased, and proceed as before, continuing the operation till all the columns have been added.

EXAMPLES.

Add the following compound numbers:

(1.)			(2.)			(3.)		
£	s.	d.	cwt.	qrs.	lbs.	℥	℈	grs.
17	13	11	2	3	27	3	1	17
13	10	2	1	1	17	2	3	19
10	17	3	4	2	26	6	1	10
£42	1s.	4d.	9 cwt.	0 qrs.	20 lbs.	12 ℥	1 ℈	6 grs.

PROOF.—The method of proof is the same as in addition of simple numbers.

(4.)			(5.)			(6)		
bu.	pks.	qts.	pks.	qts.	pts.	gals.	qts.	pts.
17	2	5	3	7	1	2	3	1
34	2	7	2	6	1	5	1	1
13	3	6	0	4	0	7	2	0
16	3	4	3	5	1	11	3	1

(7.)			(8.)			(9.)		
yds.	ft.	in.	A.	sq. ch.	sq. li.	da.	hrs.	min.
4	2	11	5	4	300	4	14	30
3	1	8	2	7	185	3	12	15
1	1	9	9	4	1230	5	4	20
6	2	1	1	8	211	6	16	18

(10.)			(11.)			(12.)		
£	s.	d.	lbs.	oz.	dwts.	mi.	rds.	yds.
18	4	9	96	10	19	2	120	4
7	11	6	4	6	16	8	72	3
9	18	9	5	10	15	2	112	4

13. Find the sum of 17*s.* 6*d.;* £3 5*s.* 8*d.;* £25 11*s.* 10½*d.;* £12 0*s.* 8*d.;* and £50 4*s.* 4½*d.*

14. Add 37 *bu.* 1 *pk.* 3 *qts.;* 41 *bu.* 2 *pks.* 5 *qts.;* 34 *bu.* 1 *pk.* 3 *qts.;* and 43 *bu.* 3 *pks.* 1 *qt.*

15. It took a carpenter 3 days to build a fence: the first day he built 4 *rds.* 4 *yds.* 1 *ft.* 4 *in.;* the second day 3 *rds.* 2 *yds.* 2 *ft.* 9 *in.;* the third day 4 *rds.* 3 *yds.* 1 *ft.* 7 *in.* What was the length of the fence?

Reducing to *half yards* and *inches* (Art. **117**.)

rds.	yds.	ft.	in.		rds.	hf. yds.	in.
4	4	1	4	=	4	8	16
3	2	2	9	=	3	5	15
4	3	1	7	=	4	7	1

Ans. 12 *rds.* 10 *hf. yds.* 14 *in.*
= 12 *rds.* 5 *yds.* 1 *ft.* 2 *in.*

If some of the numbers are fractional, they may all be reduced to decimals of the same unit and then added by the rule in Art. **92**.

16. Find the sum of 3*s.* 6*d.;* £¼; and £.875.

17. Add 3 *da.* 16 *hrs.;* ¼ of 1 *da.;* and .632 *da.*

18. Add £13 14*s.* 8*d.;* £1 7*s.* 2½*d.;* £3 13*s.* 9¼*d.;* £12 12*s.* 3¾*d.;* and £17 14*s.* 3*d.*

19. Add 4 *oz.* 15 *dwts.* 12 *grs.;* 3 *oz.* 10 *dwts.* 17 *grs.;* 11 *oz.* 14 *dwts.* 16 *grs.;* and 10 *oz.* 18 *dwts.* 29 *grs.*

20. Add 10 ℥ 7 ʒ 2 ℈ 14 *grs.;* 2 ℥ 4 ʒ 1 ℈ 18 *grs.;* 1 ℥ 3 ʒ 1 ℈ 15 *grs.;* and 3 ʒ 2 ℈ 11 *grs.*

21. Add 7 *cwt.* 3 *qrs.* 14 *lbs.;* 4 *cwt.* 2 *qrs.* 20 *lbs.;* 1 *cwt.* 1 *qr.* 10 *lbs.;* 3 *cwt.* 1 *qr.* 17 *lbs.;* 5 *cwt.* 1 *qr.* 8 *lbs.;* and 7 *cwt.* 3 *qrs.* 10 *lbs.*

22. Add 7 *wks.* 3 *da.* 11 *hrs.* ; 5 *wks.* 4 *da.* 19 *hrs.* ; 11 *wks.* 2 *da.* 13 *hrs.* ; 1 *wk.* 6 *da.* 17 *hrs.* ; and 12 *wks.* 4 *da.* 3 *hrs.*

23. Add 2 *mi.* 180 *rds.* 5 *yds.* 2 *ft.* ; 3 *mi.* 72 *rds.* 3 *yds.* 1 *ft.* ; 8 *mi.* 300 *rds.* 5½ *yds.* ; and 11 *mi.* 18 *rds.* 3 *yds.* 2 *ft.*

24. Add 8$\frac{13}{16}$ *yds.* ; 7$\frac{11}{16}$ *yds.* ; 14$\frac{23}{32}$ *yds.* ; 11$\frac{11}{32}$ *yds.* ; and 16$\frac{9}{16}$ *yds.*

25. Add 17 *bu.* 3 *pks.* 6 *qts.* ; 112 *bu.* 2 *pks.* 7 *qts.* ; 91 *bu.*

26. 29.65 *T* + 87.25 *cwt.* + 19 *lbs.* = ?

27. 1¼ *cwt.* + 17.25 *lbs.* + 49 *lbs.* = ?

28. £32.5 + 17.5*s.* + 37¼*s.* = ?

29. 47.5 *da.* + 34.2 *da.* + 7⅖ *da.* = ?

PRACTICAL PROBLEMS.

1. A merchant sent off the following quantities of butter : 47 *cwt.* 2 *qrs.* 7 *lbs.* ; 38 *cwt.* 3 *qrs.* 8 *lbs.* ; and 16 *cwt.* 2 *qrs.* 20 *lbs.* ; how much did he send off in all ?

2. A silversmith has 3 parcels of silver : the first contains 7 *lbs.* 8 *oz.* 16 *dwts.* ; the second contains 9 *lbs.* 7 *oz.* 3 *dwts.* ; and the third contains 4 *lbs.* 1 *dwt.* ; how much has he in all ?

3. A merchant sells cloth as follows : to A., 16⅞ *yds.* ; to B., 90$\frac{11}{16}$ *yds.* ; and to C., 190$\frac{1}{16}$ *yds.* ; how much does he sell to all ?

4. A man has three farms : the first contains 120 *A.* 74 *sq. rds.* ; the second contains 75 *A.* 46 *sq. rds.* ; and the third contains 97 *A.* 46 *sq. rds.* ; how much do they all contain ?

5. B. aged 14 *yrs.* 6 *mos.* goes out to service ; he lives at one place 1 *yr.* 9 *mos.*, at another place 2 *yrs.* 5 *mos.*, and at a third place 3 *yrs.* 9 *mos.* ; how old is he then ?

6. A man spent 17.25 *francs* for a vest, 62.17 *fr.* for a coat, and 38.29 *fr.* for a pair of boots; how many dollars did he spend in all ?

7. A man sold 4 cheeses: the first weighed 9.25 *kilog.*, the second 10.14 *kilog.*, the third 11.16 *kilog.*, and the fourth 10.77 *kilog.*; how many pounds did they weigh ?

8. In a farm there are 5 fields: the first contains 18 *A.* 8 *sq. ch.*, the second 12 *A.* 3 *sq. ch.*, the third 9 *A.* 4 *sq. ch.*, the fourth 11 *A.*, and the fifth 16 *A.* 2 *sq. ch.;* what is the content of the farm ?

9. How many yards in 4 pieces of cloth measuring as follows: 30⅝ *yds.*, 27¾ *yds.*, 39 1/16 *yds.*, and 37½ *yds.?*

10. A man bought 3 loads of wood: the first contained 1 *C.* 17 *cu. ft.*, the second 1 *C.* 115 *cu. ft.*, and the third 1 *C.* 2 *C. ft.;* how much wood did he buy ?

REVIEW QUESTIONS.

(**149.**) What is addition of compound numbers? (**150.**) What is the rule for addition of compound numbers?

IV. SUBTRACTION OF COMPOUND NUMBERS.

DEFINITION.

151. **Subtraction of Compound Numbers** is the operation of finding the **Difference** of two numbers of the same kind.

MENTAL EXERCISES.

1. What is the difference of 14 *qts.* and 9 *qts.?* of 7 *pks.* and 4 *pks.?* of 18 *yds.* and 11 *yds.?* of 30 *cts.* and 17 *cts.?* of 25 *rds.* and 14 *rds.?* of 16 *mi.* and 11 *mi.?*

2. What is the difference of 31*d*. and 16*d*.? How many *shillings* in 31*d*.? in 16*d*.? In the difference between 31*d*. and 16*d*.? What is the difference between 2*s*. 7*d*. and 1*s*. 4*d*.? 2*s*. 7*d*. — 1*s*. 4*d*. = ?

3. How many *yards* in 26 *ft*.? in 16 *ft*.? What then is the difference between 8 *yds*. 2 *ft*. and 5 *yds*. 1 *ft*.? between 8 *yds*. 1 *ft*. and 6 *yds*. 2 *ft*.? between 7 *gals*. 1 *qt*. and 5 *gals*. 3 *qts*.?

NOTE.—The principles used in *subtraction of compound numbers* are the same as those used in *subtraction of simple numbers*.

OPERATION OF SUBTRACTION OF COMPOUND NUMBERS.

152. Let it be required to find the difference between £9 4*s*. 3*d*. and £2 18*s*. 6*d*.:

EXPLANATION.—We write the subtrahend under the minuend so that units of the same denomination shall stand in the same column. Beginning at the lowest denomination, we see that 6*d*. cannot be taken from 3*d*.; we therefore add 12*d*. to 3*d*., which gives 15*d*., and then subtract 6*d*. from the sum; the remainder, 9*d*., we set down, and to compensate for the 12*d*. added to the minuend, we add its equal, 1*s*., to the next column of the minuend. The sum, 19*s*., being greater than 4*s*., we add 20*s*. to the latter and subtract 19*s*. from the sum; the remainder, 5*s*., we set down and as before carry forward 20*s*., or its equal £1, and add it to the minuend, giving £3; this taken from £9 leaves £6: hence, the required remainder is £6 5*s*. 9*d*.

OPERATION.

£	*s.*	*d.*
9	4	3
2	18	6
£6	5*s.*	9*d.*

In like manner we may treat all similar cases; hence, the following

RULE.

I. Write the subtrahend under the minuend so that units of the same denomination shall stand in the same column.

II. Subtract each number in the lower line from the one above it and write the remainder in the line below.

III. If any number in the lower line is greater than the one above it, increase the latter by as many units as make one of the next higher denomination, perform the subtraction and then add 1 unit to the next number in the lower line.

EXAMPLES.

Perform the following indicated subtractions:

	(1.)			(2.)			(3.)		
	£	s.	d.	lbs.	oz.	dwts.	bu.	pks.	qts.
From	14	14	3	6	11	14	65	1	7
Take	9	17	1	2	3	16	14	3	4

Rem. £4 17s. 2d.   4 lbs. 7 oz. 18 dwts.   50 bu. 2 pks. 3 qts.

Proof.—The method of proof is the same as for subtraction of simple numbers.

(4.)			(5.)			(6.)		
cwt.	qrs.	lbs.	hhds.	gals.	qts.	yds.	ft.	in.
7	3	13	112	23	1	4	2	11
5	1	15	75	37	1	2	2	9

(7.)		(8.)			(9.)				
acres.	sq. rds.				℔	℥	ʒ	℈	grs.
29	50	23°	45'	54"	35	7	3	1	14
24	65	7°	49'	57"	17	10	6	1	18

10. From 4 rds. 2 yds. 1 ft. 9 in. subtract 2 rds. 3 yds. 1 ft. 11 in.

OPERATION.

rds.	yds.	ft.	in.		rds.	hf. yds.	in.
4	2	1	9	=	4	5	3
2	3	1	11	=	2	7	5

$$\text{1 } rd. \quad 8 \text{ } hf.rds. \text{ } 16 \text{ } in.$$
$$= \text{1 } rd. \quad 4 \text{ } yds. \text{ } 1 \text{ } ft. \text{ } 4 \text{ } in. \text{ } Ans.$$

11. From 12 *rds.* 2 *yds.* 2 *ft.* 1 *in.* subtract 3 *rds.* 3 *yds.* 2 *ft.* 10 *in.*

12. From 8 *rds.* 1⅖ *yds.* subtract 3 *rds.* 3 *yds.* 2 *ft.* 6 *in.*

NOTE.—To write a date as a compound number, we first write the number of the current year, then the number of the current month, counted from the beginning of the year (Art. **114**), and then the number of the day. Thus, July 7th, 1839, is written 1839 *yrs.* 7 *mos.* 7 *da.* In computing the difference of two dates, a month is to be counted equal to 30 days.

13. What is the difference of time between October 16th, 1869, and Aug. 2d, 1873?

OPERATION.

August 2d, 1873 . . . 1873 *yrs.* 8 *mos.* 2 *da.*
October 16th, 1869 . . 1869 *yrs.* 10 *mos.* 16 *da.*

$$\text{3 } yrs. \quad 9 \text{ } mos. \quad 16 \text{ } da. \quad Ans.$$

14. How long from Sept. 25, 1871, to July 4, 1876?

15. How long from July 7, 1815, to Nov. 1, 1873?

16. How long from May 13, 1816, to June 25, 1859?

17. What is the difference between 22 *hrs.* 17 *min.* 4 *sec.* and 14 *hrs.* 9 *min.* 51 *sec.*?

18. What is the difference between £1.5 and 7*s.* 6*d.*?

19. From $\frac{4}{15}$ of 1 *hhd.* subtract $\frac{4}{9}$ of 1 *qt.*

20. From 3.107 *kilog.* subtract 331.2 *grams.*

21. From 16 *da.* 21 *hrs.* 42 *min.* 13 *sec.* subtract 12 *da.* 22 *hrs.* 58 *min.* 39 *sec.*

22. From 7 *T.* 14 *cwt.* 3 *qrs.* 19 *lbs.* subtract 3 *T.* 18 *cwt.* 1 *qr.* 4 *lbs.*

23. From 14 *lbs.* 1 *oz.* 3 *dwts.* 18 *grs.* subtract 9 *lbs.* 0 *oz.* 16 *dwts.* 5 *grs.*

24. From 4,306 *gals.* 1 *qt.* subtract 3,621 *gals.* 2 *qts.* 1 *pt.*

25. From 110 *bu.* 1 *pk.* 2 *qts.* subtract 94 *bu.* 3 *pks.* 7 *qts.*

PRACTICAL PROBLEMS.

1. A merchant bought a piece of cloth for £22 10*s.* and sold one half of it for £14 18*s.* ; for what must he sell the rest to make £7 14*s.* 3*d.?*   *Ans.* £15 6*s.* 3*d.*

2. A farm contains 273 *A.* 1 *R.* 5 *sq. rds.*, but only 111 *A.* 2 *R.* 38 *sq. rds.* was capable of tillage; how much of it was incapable of tillage ?

3. From a piece of cloth containing 39$\frac{11}{16}$ *yds.*, there was cut off at one time 3$\frac{4}{5}$ *yds.* and at another time 4$\frac{1}{4}$ *yds.;* how much remained in the piece ?

4. A merchant has 183 *cwt.* 24 *lbs.* of butter, of which he ships 78 *cwt.* 3 *qrs.* 14 *lbs.;* how much remains ?

5. How long from Jan. 20, 1873, to Nov. 14, 1875 ?

6. A man was born Jan. 10, 1803, and died Sept. 21, 1875; what was his age at the time of his death ?

7. The revolutionary war began April 19, 1775, and ended Jan. 20, 1783; how long did it last ?

8. How long from the discovery of America, Oct. 11, 1492, to the declaration of independence, July 4, 1776?

9. From a pile of wood containing 11 *C.* 4 *C. ft.*, there was sold 4 *C.* 5 *C. ft.* 12 *cu. ft.;* how much remained?

10. The latitude of Albany is 42° 39' 3″ N., and that of St. Petersburg is 59° 56' N. ; what is the difference ?

Ans. 17° 16' 57″.

NOTE.—The **Latitude** of a place is its angular distance from the equator. If the place is north of the equator its latitude is marked N., if south, its latitude is marked S. If the latitude of two places are both *north* or both *south*, their *difference of latitude* is found by subtracting the less from the greater ; if the latitude of one place is north and the other south, their *difference of latitude* is found by adding the latitudes of both.

11. The latitude of New York is 40° 42' 45" N., that of the Cape of Good Hope is 34° 22' S ; what is the difference ? *Ans.* 75° 4' 45".

12. The latitude of St. Augustin is 29° 48' 30" N., and that of Gibraltar is 36° 7' N. ; what is the difference ?

13. The longitude of New York is 74° 3' W., and that of San Francisco is 122° 26' 45" W. ; what is their difference of longitude ? *Ans.* 48° 23' 45".

NOTE.—Longitudes are reckoned both *east* and *west* from some assumed meridian, usually that of Greenwich, England. The method of finding *difference of longitude* is the same as for finding difference of latitude.

14. The longitude of Berlin is 13° 24' E., and that of Washington is 77° 0' 15" W. ; what is the difference ?

15. The longitude of Charleston is 79° 55' 38" W., and that of Boston is 71° 3' 30" W. ; what is the difference ?

16. A farmer has 147 *bu.* 1 *pk.* of oats ; he puts 49 *bu.* 3 *pks.* in one bin, 27 *bu.* 1 *pk.* in a second bin, 32 *bu.* 3 *pks.* in a third bin, and the rest in a fourth bin ; how many does he put in the fourth bin ?

REVIEW QUESTIONS.

(151.) What is subtraction of compound numbers? **(152.)** Give the rule. How proved ? How are dates written ? What is the latitude of a place ? How reckoned ? What is the method of finding difference of latitude ? Difference of longitude ?

V. MULTIPLICATION OF COMPOUND NUMBERS.

DEFINITION.

153. **Multiplication of Compound Numbers** is the operation of taking a compound number as many times as there are units in an abstract number.

MENTAL EXERCISES.

1. How many *inches* are 7 times 8 inches? What is the product of 11 *in.* by 9 ? of 7*d.* by 11 ? of 6 *oz.* by 14 ? of 16 *yds.* by 8 ?

2. What is the product of 8*d.* by 11 ? How many *shillings* in the product and how many pence remain ? What is the product of 2 *ft.* by 16 ? How many *yards* and *feet* in the product ? What is the product of 7 *qts.* by 13 in *pecks* and *quarts* ?

3. What is the product of 15 *in.* by 9 ? How many *feet* and *inches* in 15 *in.?* in 9 times 15 *in.?* What then is the product of 1 *ft.* 3 *in.* by 9 ? What is the product of 4 *bu.* 3 *pks.* by 11 ? of 1*s.* 7*d.* by 8 ?

NOTE.—The principles used in *multiplication of compound numbers* are the same as those used in *multiplication of simple numbers*.

OPERATION OF MULTIPLICATION OF COMPOUND NUMBERS.

154. Let it be required to multiply £4 2*s.* 5*d.* by 16 :

EXPLANATION.—Having written the multiplier under the multiplicand, we multiply 5*d.* by 16, which gives 80*d.*, or 6*s.* 8*d.* ; setting down 8*d.*, we carry 6*s.* to the next column. We then multiply 2*s.* by 16 and add 6*s.* to the product, which gives 38*s.*, or £1 18*s.* ; setting down 18*s.*, we carry £1 to the next column. Finally, we

OPERATION.

£	*s.*	*d.*
4	2	5
		16
£65	18*s.*	8*d.*

multiply £4 by 16 and add £1 to the product, which gives £65. Hence, the required product is £65 18s. 8d.

In like manner we may treat all similar cases; hence, the following

RULE.

I. Multiply the units of the lowest denomination of the multiplicand by the multiplier, and divide the product by the number of the scale that connects this denomination with the next higher one; set down the remainder and carry the quotient to the next column.

II. Multiply the units of the next higher denomination by the multiplier, add the units brought forward, and proceed as before, continuing the operation till all the parts of the given number have been multiplied.

EXAMPLES.

	(1.)			(2.)			
	£	s.	d.	cwt.	qrs.	lbs.	oz.
Multiplicand.	17	15	9	8	3	1	9
Multiplier...			6				7
Product....	£106	14s.	6d.	61 cwt.	1 qr.	10 lbs.	15 oz.

(3.)

mi.	rds.	yds.	ft.	in.	mi.	rds.	hf. yds.	in.
9	110	4	2	6	= 9	110	9	12
				9				9

84 mi.   37 rds. 10 hf. yds. 0 in.

= 84 mi. 37 rds. 5 yds. *Ans.*

Multiply

4. 5 *cwt.* 2 *qrs.* by 7.

Multiply

5. $8.75 by 24.5.

6. 65.35 *fr.* by 46.

7. 5.84 *kilog.* by 12.

8. 15 *yds.* 1 *ft.* by 21.

9. 123.25 *m.* by 15.

10. 6 *wks.* 3 *da.* by 13.

11. 17 *gals.* 2 *qts.* by 18.

12. 15 *lbs.* 3 *oz.* by 16.

13. 13 *yds.* 2⅓ *ft.* by 9.

14. £⅔ by 17.5.

15. 10 *A.* 1 *R.* by 11.

16. 3 *hrs.* 15¼ *min.* by 24.

17. 5 *cwt.* 2⅖ *qrs.* by 24.

18. £3 14⅝*s.* by 33.

19. 18 *T.* 5 *cwt.* by 127.

20. 4 *yds.* 2 *ft.* by 18.75.

21. 7 *gals.* 3 *qts.* by 14.72.

22. How many *square feet* in a rectangle whose length is 17 *ft.* 5 *in.*, and whose breadth is 3 *ft.* 9 *in.?*

$$Ans. \ 17.417 \times 3.75 = 65.314 \, sq. \, ft.$$

NOTE.—In examples like the above, we reduce both dimensions to decimals of a foot, as explained in Art. **148**; we then multiply the number of feet in the length by the number of feet in the breadth; the product will be the number of *square feet* in the area (Art. **122**).

To secure uniformity in this and all similar cases, let the student carry decimals to *three places*, applying the *rule for approximation*, given in Art. **90,** at *each step* of the operation.

23. How many *square yards* in the floor of a room whose length is 22 *ft.* 4 *in.*, and whose breadth is 16 *ft.* 9 *in.?*

SOLUTION.—Because 22 *ft.* 4 *in.* = 7.444 *yds.*, and 16 *ft.* 9 *in.* = 5.583 *yds.*, we have 7.444 × 5.583 = 41.561 *sq. yds.* *Ans.*

24. How many *sq. ft.* in a floor 16 *ft.* 6 *in.* long and 13 *ft.* 9 *in.* wide?

SOLUTION.—16.5 *ft.* × 13.75 *ft.* = 226⅞ *sq. ft.* *Ans.*

25. How many *sq. yds.* in a plot 102 *yds.* long and 94 *yds.* 2 *ft.* wide?

26. How many *sq. rds.* in a field 42½ *rds.* long and 7¼ *rds.* wide?

27. How many cubic feet in a bin 7 *ft.* 2 *in.* long, 3 *ft.* 4 *in.* wide, and 2 *ft.* 9 *in.* deep ?

Ans. 7.167 × 3.333 × 2.75 = 65.682 cu. ft.

EXPLANATION.—Here we reduce each dimension to the required linear unit, and then find the continued product of the corresponding numbers. First, we find 7.167 × 3.333 = 23.888; we then find 23.888 × 2.75 = 65.682.

28. How many *cubic yards* in rectangular volume of earth 27 *yds.* 1 *ft.* long, 13 *yds.* 2 *ft.* wide, and 2 *yds.* high ?

29. How many *cubic feet* in a hewn stick of timber 27 *ft.* long, 1 *ft.* 2 *in.* wide, and 9 *in.* thick ?

30. How many *cubic feet* in a room 12½ *ft.* long, 10½ *ft.* wide, and 7¾ *ft.* high ? How many *square feet* in the floor ?

31. What is the product of 17 *ft.* 9 *in.* by 14 *ft.* 6 *in.*

PRACTICAL PROBLEMS.

1. If a gentleman spends £1 7s. 6d. a day, how much will he spend in 365 days ?

2. What is the length of 36 pieces of telegraph wire, the length of each piece being 26 *mi.* 1,125 *yds.?*

3. What is the weight of 37 parcels of silver, averaging 12 *lbs.* 2 *oz.* 15 *dwts.* 6 *grs.* each ?

4. How many yards of cloth in 27 bales, each bale containing 15 pieces, and each piece 15⅛ *yds.?*

5. In 7 loads of wood, each containing 2 *C.* 3 *C. ft.,* how many cords ? How many *cubic feet?*

6. What is the weight of 1½ *dozen* silver forks, each fork weighing 2 *oz.* 3 *dwts.?*

7. What is the weight of 7 *hhds.* of sugar, each hogshead weighing 5 *cwt.* 1 *qr.* 14 *lbs.?*

8. How far can a man travel in 5 *days*, at the rate of 7½ *mi.* a day?

9. How much land is there in 9 fields, each containing 8 *A.* 2 *R.* 30 *sq. rds.?*

10. What do 54 sheep cost at 15*s.* 3*d.* each?

11. How many *yds.* in 12 pieces, each containing 19¾ *yds.?*

12. What is the cost of 52⅜ *yds.* of cloth at $3.50 a yard?

13. How many *yds.* in 6½ pieces of muslin, each containing 39¼ *yds.?*

14. What is the weight of 15 loads of hay, each weighing 1 *T.* 3 *cwts.?*

15. How many *sq. yds.* of flagging will be required to flag a court 117 *ft.* long and 98 *ft.* 3 *in.* wide?

16. How many *square feet* of boards will be required to make a floor 22 *ft.* 6 *in.* long and 19 *ft.* 9 *in.* broad?

17. A room is 18½ *ft.* long, 13¼ *ft.* wide, and 9¼ *ft.* high; how many *square yards* of carpeting will it require to cover the floor? How many *square feet* of kalsomining in the ceiling? How many *square yards* of paper will it take to cover the walls, the doors and windows not being taken into account?

18. A rectangular block of stone is 6 *ft.* long 3 *ft.* wide, and 2½ *ft.* thick; what is its weight, if each cubic foot weighs 156 *lbs.?*

REVIEW QUESTIONS.

(**153.**) What is multiplication of compound numbers? (**154.**) Give rule for multiplication of compound numbers? How do you find the contents of a rectangular area? Of a rectangular solid?

VI. DIVISION OF COMPOUND NUMBERS.

DEFINITION.

155. **Division of Compound Numbers** is the operation of dividing a compound number by an abstract number, or by a similar denominate number.

MENTAL EXERCISES.

1. If 35 nuts are divided equally among 7 boys, how many will each receive? What is the quotient of 35 *nuts* by 7? Of 18*d.* by 6? Of 27 *horses* by 9? Of 48 *yds.* by 6? Of 156 *bu.* by 13? Of 156 by 13?

2. If 27 *marbles* are divided in piles, each containing 9 *marbles*, how many piles will there be? What is the quotient of 27 *marbles* by 9 *marbles?* Of 49 *rods* by 7 *rods?* Of 96*s.* by 12*s.?* Of 180*d.* by 15*d.?* Of 180 by 15?

3. What is the quotient of 153*d.* by 9? How many *shillings* in 153*d.*, and how many *pence* remain? How many *shillings* and *pence* in 17*d.?* What then is the quotient of 12*s.* 9*d.* by 9? What is the unit of 153*d.?* Of 9*d.?* Of 17*d.?*

4. What is the quotient of 84 *yds.* by 12 *yds.?* What is the unit of 84 *yds.?* Of 12 *yds.?* Of 7? Is there any difference between the quotient of 57 *lbs.* by 19 *lbs.* and of 57 by 19?

5. A floor contains 40 *square yards*, and its length is 5 *yards;* what is its breadth? What is the quotient of 4 *square yards* by 5 lineal yards?

NOTE.—The principles used in division of compound numbers are the same as those used in division of simple numbers.

OPERATION OF DIVISION OF COMPOUND NUMBERS.

156. Let it be required to divide £65 18s. 8d. by 16:

EXPLANATION.—Dividing £65 by 16, we find a quotient £4, and a remainder £1. Reducing £1 to shillings and adding 18s., we have 38s., which we take for a new dividend ; dividing 38s. by 16, we find a quotient 2s. and a remainder 6s. Reducing 6s. to pence and adding 8d., we have 80d., which we take for a new dividend ; dividing 80d. by 16, we find the quotient 5d. and a remainder 0. The required quotient is therefore £4 2s. 5d.

OPERATION.

$$
\begin{array}{r}
£ \quad s. \quad d. \\
16\ )\ 65 \quad 18 \quad 8\ (\ £4\ 2s.\ 5d. \\
\underline{64} \\
£1\ \ldots\ \text{1st } rem. \\
20 \\
\overline{38s.} \\
32 \\
\overline{6s.\ \ldots\ \text{2d } rem.} \\
12 \\
\overline{80d.} \\
80 \\
\overline{}
\end{array}
$$

In like manner we may treat all similar cases ; hence, the following

RULE.

I. Divide the units of the highest denomination in the dividend by the divisor and write the quotient as a part of the required quotient.

II. Reduce the remainder to the next lower denomination, and to the result add the units of that denomination, for a new dividend, and proceed as before.

III. Continue this operation till the division is completed.

EXAMPLES.

Perform the following indicated divisions:

(1.)

7) 37 *bu.* 3 *pks.* 7 *qts.*

Quotient, 5 *bu.* 1 *pk.* 5$\frac{4}{7}$ *qts.*

(2.)

9) 1 *T.* 19 *cwt.* 2 *qrs.* 12 *lbs.*

4 *cwt.* 1 *qr.* 15$\frac{2}{3}$ *lbs.*

3. Divide 17 *cwt.* 0 *qrs.* 2 *lbs.* 6 *oz.* by 7.

4. Divide 228 *T.* 18 *cwt.* 3 *qrs.* 13 *lbs.* 12 *oz.* by 11.

5. Divide 9 *hhds.* 28 *gals.* 2 *qts.* by 49 *gals.* 2 *qts.* 1 *pt.*

EXPLANATION.—The dividend is equal to 4764 *pts.* and the divisor is equal to 397 *pts.* ; hence, the quotient is equal to 4764 *pts.* ÷ 397 *pts.* = 12. *Ans.* All similar cases may be solved by the following

RULE.

Reduce both numbers to the same denomination and divide as in simple numbers.

NOTE.—If the divisor is abstract, the quotient is similar to the dividend ; if the divisor is similar to the dividend, the quotient is abstract.

6. Divide 17 *lea.* 1 *mi.* 4 *fur.* 21 *rds.* by 21.

7. Divide 25 *bu.* 3 *pks.* 4 *qts.* by 9.

Perform the following indicated divisions:

8. £21 11s. 3d. by 15.	20. 129½ *bu.* by 8 *bu.* 3 *qts.*
9. 15 *lbs.* 3 *oz.* 12 *dwts.*÷12.	21. 58650 *m.* ÷ 3.45 *kilom.*
10. 39 *kilog.* by 7.5.	22. 56° 45′ ÷ 15°.
11. 3 *mi.* by 22 *ft.* 6 *in.*	23. 78° 16′ 15″ ÷ 15.
12. 1361 *mi.* 188 *rds.*÷ 28.	24. 69° 4′ 45″ ÷ 15.
13. 117.9 *fr.* by 131.	25. 2 *gals.* 3 *qts.*÷ 3 *qts.*
14. 203 *hectol.* by 58.	26. 3 *T.* 16 *cwt.*÷ 19 *lbs.*
15. £30. 7s. 1¼d. by 7.	27. 12 *hrs.* 48 *min.*÷ 16 *sec.*
16. 9 *hhds.* 28½ *gals.* by 12.	28. 15.75 *yds.*÷ 3.5 *yds.*
17. 3 *mi.* by 2 *ft.* 3 *in.*	29. 24 *rds.* 3 *yds.*÷ 27 *ft.*
18. 186.02 *fr.* by 131.	30. 48° 4′ 30″ ÷ 45°.
19. 39.06 *kilog.* by 93.	31. £6 1s. 6d. ÷ 27.

NOTE.—When the divisor is composite, we may divide by each of its factors in succession. In the last example, if we divide by 3, we find for a quotient £2 0s. 6d. ; and this result divided by 9 gives 4s. 6d. *Ans.*

32. Divide £37 14s. by 24, that is, by 4×6.

33. Divide 178 *lbs.* 9 *oz.* 14 *dwts.* 16 *gr.* by 77.

34. Divide 147 *bu.* 3 *pks.* 4 *qts.* by 5 *bu.*

35. Divide 2 *T.* 5 *cwt.* 24 *lbs.* by 12 *lbs.*

PRACTICAL PROBLEMS.—Miscellaneous.

1. If 7 calves cost £15 1s., what is the cost of 1 calf? of 5 calves? of 13 calves?

2. If a man can walk 38¼ miles in 11 hours, how far can he walk in 3 hours? in 9 hours?

3. How many times can a 3-quart measure be filled from a cask of wine containing 26 *gals.* 1 *qt.?*

4. How many *yds.* of cambric ⅞ *yd.* wide will it take to line 14 *yds.* of cloth 1¼ *yds.* wide?

5. A garden containing 1,154¼ sq. *yds.* is 40¼ *yds.* long; how wide is it?

6. A stick of hewn timber containing 30 *cu. ft.* is 32 *ft.* long and 1 *ft.* 3 *in.* wide; what is its thickness?

7. What is the cost of 1½ *mi.* of iron pipe at 12½ *cts.* a foot?

8. How many panes of glass, each 12 *in.* by 15 *in.*, in a box of glass containing 100 *sq. ft.?*

9. If 4 *qts.* of salt cost 21 *cts.*, what is the cost of 4 *bu.* 3 *pks.?*

10. What will it cost to carpet a floor 18 *ft.* long and 15¾ *ft.* wide, with carpeting ¾ *yd.* wide and costing $1¾ a yard?

11. A room is 24 *ft.* long, 19½ *ft.* wide, and 10½ *ft.* high; what will it cost to plaster the 4 sides of the room at the rate of 30 *cts.* a *sq. yd.* after deducting ⅕ for doors and windows?

12. What will it cost to plaster and kalsomine the ceiling of the room described in Problem 11, at 40 *cts.* a square yard?

13. What will it cost to carpet the same room with carpeting 1 *yd.* wide at $2.50 a yard?

14. A walk 4½ *ft.* wide and 180 *ft.* long is to be flagged with stones 18 *in.* long and 9 *in.* wide; how many will it take?

15. How many rolls of paper, each 19 *ft.* long and 1 *ft.* 6 *in.* wide, will it take to paper the sides of a room 18 *ft.* long, 15 *ft.* wide, and 9½ *ft.* high, no allowance being made for doors or windows?

16. A man divides 43 *A.* 3 *R.* 20 *sq. rds.* into 18 equal building lots; how much in each lot?

17. A farm of 154 *A.* 20 *sq. rds.* is laid out in lots containing 12 *A.* 3 *R.* 15 *sq. rds.* each; how many lots are there?

18. A merchant buys 10 *cwt.* 2 *qrs.* of sugar for $131.25, for what must he sell it to make 1¼ *cts.* a pound?

19. A speculator bought 4 city lots, each 25 by 100 *feet*, at $1¼ a square foot, and sold them again for $20,470; what did he gain on each lot?

20. If 20 bricks will build 1 cubic foot of wall, how many bricks will be required to build a wall 70 *ft.* long, 1½ *ft.* thick, and 4 *ft.* high?

21. How many sheets of tin, 18 *in.* long and 15 *in.* wide, will it take to cover the roof of a house 40 feet long, the rafters on each side being 18 feet long?

22. What will it cost to floor a room 17½ *ft.* long and 16 *ft.* wide, at the rate of $1.10 per square yard?

23. If it costs $62.50 to lay 50 *sq. yds.* of flooring, what will it cost to lay a floor 33 *ft.* long and 18 *ft.* wide ?

24. How many spoons, each weighing 2 *oz.* 10 *dwts.*, can be made from a bar of silver weighing 11 *lbs.* 3 *oz.?*

25. The circumference of the fore wheel of a carriage is 13 *ft.* 9 *in.* and that of the hind wheel 16 *ft.* 6 *in.;* how many times more will the fore wheel turn than the hind one in a journey of 30 miles ?

26. If 160 bushels of oysters cost £75 17s. 4d., what does 1 bushel cost ?

27. A truckman carried 117 *C.* 110 *cu. ft.* of wood in 100 equal loads; how much did he carry at each load ?

28. If 3 *yds.* of cloth cost £4 16s. 6d., how much does 1 yard cost, and how much does 12 *yds.* cost ?

29. A person's yearly income is 14,636.80 *fr.* ; of this he gives in charity 2,500 *fr.;* his weekly bills are 149.15 *fr.* each ; and the rest he spends in traveling ; how much does he spend per week in traveling ?

30. The average speed of a railway train is 4 *myriameters* per hour; how long will it take to travel from Paris to Boulogne, a distance of 272 *kilom.?*

31. Bought the following articles: 27¼ *meters* of linen at 3.50 *fr.* per meter ; 6¼ *meters* of velvet at 17 *fr.* per meter ; 29 *meters* of brocade at 16.75 *fr.* per meter ; 28¾ *meters* of merino at 4 *fr.* per meter; 1½ *doz. pairs* of socks at 3.05 *fr.* per pair; and 7 *pairs* of gloves at 42 *fr.* a dozen pairs ; what was the amount of the bill ?

32. If a person takes 108 steps a minute, each step being 30 inches, how far can he walk in 2 *hrs.* 30 *min.?* How far can he walk in 3 *hrs.* 10 *min.?*

33. How many yards of cloth will it take to clothe a company of 48 men, if 39⅖ *yds.* will clothe 7 men ?

34. The longitude of Philadelphia is 75° 9′ 23″ W.; that of San Francisco is 122° 24′ 39″ W.: what is the difference ?

35. The latitude of Dublin is 53° 23′ N.; that of Santiago is 33° 26′ 26″ S.: what is the difference ?

APPLICATION TO LONGITUDE AND TIME.

157. From the explanation of longitude in Art. **137**, it appears that the sun comes to the meridian of a place whose longitude is 15° W., 1 hour later than it comes to the standard meridian. Hence, the local times at the two meridians differ by 1 *hour;* at places whose longitudes differ by 30° the difference of local times is 2 *hours;* and so on, according to the Table in Art. **137.**

EXAMPLES.

1. Let it be required to find the difference of time at two places where difference of longitude is 74° 1′.

EXPLANATION. — Dividing 74 by 15, we find a quotient 4, and a remainder 14°; we call the quotient 4 *hrs.:* reducing 14° to *minutes of arc* and adding 1′, we

OPERATION.

$$15\,\overline{)\,74°\ \ 1′}$$
$$4\,hrs.\ 56\,min.\ 4\,sec.$$

have 841′, which divided by 15 gives a quotient 56, and a remainder 1′; we call the quotient *minutes of time,* 56 *min.;* reducing 1′ to *seconds of arc,* we have 60″, which divided by 15 gives 4; this we call *seconds of time,* 4 *sec.* Hence, the required difference of time is 4 *hrs.* 56 *min.* 4 *sec.*

By reversing the process just explained, we can find the difference of longitude of two places when we know the difference of their local times.

Hence, the following

R U L E .

1°. Divide the difference of longitude in arc by 15, changing degrees, minutes, and seconds of arc to hours, minutes, and seconds of time; the result will be the difference of time.

2°. Multiply the difference of time by 15, changing hours, minutes, and seconds of time to degrees, minutes, and seconds of arc; the result will be the difference of longitude in arc.

Reduce	Reduce
2. 17° 24′ 15″ to time.	8. 3 *hrs.* 4 *min.* 6 *sec.* to arc.
3. 54° 18′ 45″ "	9. 2 *hrs.* 9 *min.* 18 *sec.* "
4. 118° 23′ 30″ "	10. 5 *hrs.* 14 *min.* 23 *sec.* "
5. 21° 47′ 45″ "	11. 9 *hrs.* 17 *min.* 10 *sec.* "
6. 79° 40′ 15″ "	12. 1 *hr.* 18 *min.* 36 *sec.* "
7. 38° 38′ 45″ "	13. 6 *hrs.* 29 *min.* 4 *sec.* "

14. The difference of longitude between Rochester, N.Y., and San Diego, Cal., is 39° 22′ 22″; what is their difference of time?

15. The difference of time between Liverpool and West Point is 4 *hrs.* 43 *min.* 48 *sec.;* what is their difference of longitude?

REVIEW QUESTIONS.

(**155.**) What is division of compound numbers? (**156.**) Give the rule when the dividend is denominate and the divisor abstract. When the dividend and divisor are similar denominate numbers. (**157.**) How do you reduce difference of longitude to difference of time? Difference of time to difference of longitude?

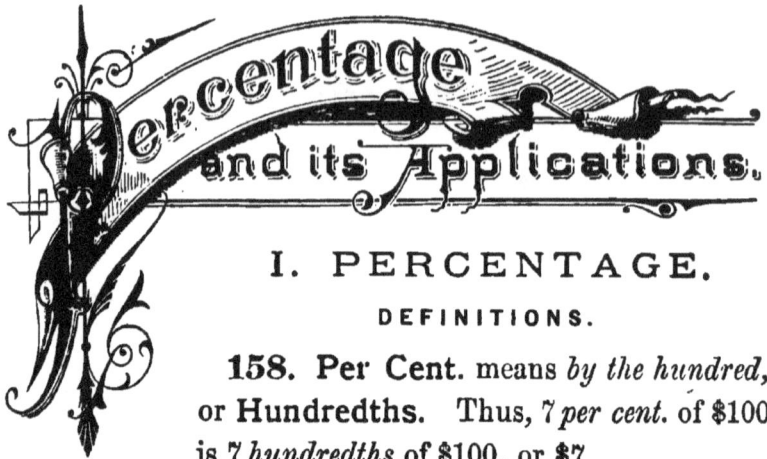

I. PERCENTAGE.

DEFINITIONS.

158. **Per Cent.** means *by the hundred,* or **Hundredths.** Thus, 7 *per cent.* of $100 is 7 *hundredths* of $100, or $7.

The **Sign**, %, is read *per cent.* Thus, 7% of $100 is read 7 *per cent.* of $100.

159. The **Rate Per Cent.**, or simply the *rate*, is the *number of hundredths* taken. Thus, in the expression 7% of $100, the *rate* is 7 *hundredths*, or .07.

NOTE.—*Rate per cent.* may be expressed in any of the ways shown in the following

TABLE OF EQUIVALENTS.

7   *per cent.*, or   7 %, is equivalent to $\frac{7}{100}$, or to .07.
12½ *per cent.*, or 12½%, "   " $\frac{12.5}{100}$, "   .125.
100   *per cent.*, or 100 %, "   " $\frac{100}{100}$, "   1.
125   *per cent.*, or 125 %, "   " $\frac{125}{100}$, "   1.25.
142¾ *per cent.*, or 142¾%, "   " $\frac{142.75}{100}$, "   1.4275.
         etc.,   etc.,   etc.,   etc.

160. Percentage is some *per cent.* of a given number. Thus, $7 is the *percentage* on $100 when the rate is 7 *per cent.*

NOTE.—The general term *percentage* is applied to all operations in which the computation is made by hundredths.

161. The **Base** is the number on which percentage is reckoned. Thus, in the expression 7% of $100, the *base* is $100.

MENTAL EXERCISES.

1. How many per cent. is 4 *hundredths?* .09 ? .74 ? .12¼? How many hundredths is 11% ? 15% ? 37½% ? 150% ?

2. What is 4 *hundredths* of $100 ? What is 4% of $100 ? 5% of $80 ? 9% of 16 *lbs.?* 12% of 18 *ft.?* 11% of $14 ?

3. What is 12% of 20 *yds.?* What is the *base?* The *rate per cent.?* The *percentage?* What is 37½% of $100 ? What is the base ? The rate ? The percentage ? If you multiply the base by the rate, what is the product ?

ADDITIONAL DEFINITIONS.

162. The **Amount** is the base *increased* by the percentage. Thus, the *amount* of $100 increased by 8%, is $100 + $8, or $108.

163. The **Difference** is the base *diminished* by the percentage. Thus, the *difference* of $100 diminished by 8%, is $100 — $8, or $92.

NOTE.—Both *amount* and *difference* are *percentages.* Thus, the *amount* of $100 increased by 8 % is 108 % of $100 ; and the *difference* of $100 diminished by 8 % is 92 % of $100.

MENTAL EXERCISES.

1. What is 5% of 40 *lbs.?* If 40 *lbs.* is increased by 5% of 40 *lbs.*, what is the amount ? What per cent. of 40 *lbs.* is the amount ? What is the amount of $80 increased by 15% of $80 ? What per cent of $80 is $92 ?

2. What is 7% of $60? If $60 is diminished by 7% of $60, what is the difference ? How many per cent. of $60

is the difference? What is the difference of 30 *yds.* and 20% of 30 *yds.*? What per cent. of 30 *yds.* is 24 *yds.*?

3. The base is $20 and the rate is 7%, what is the amount? The base is 15 *lbs.* and the rate 6%, what is the difference? The base is 20 *yds.* and the rate 10%, what is the amount and what is the difference?

4. A man had 30 chickens, but 20% of them were destroyed by a fox; how many per cent. were left? How many chickens were destroyed? How many chickens were left? What is 60% of 30 chickens?

PRINCIPLES.

164. From what precedes we have the following principles:

1°. *The percentage is equal to the base multiplied by the rate.*

2°. *The amount is equal to the base multiplied by 1 plus the rate.*

3°. *The difference is equal to the base multiplied by 1 minus the rate.*

Because either of two factors is equal to their product divided by the other, we have the following principles:

4°. *The rate is equal to the percentage divided by the base.*

5°. *The base is equal to the percentage divided by the rate; to the amount divided by 1 plus the rate; or, to the difference divided by 1 minus the rate.*

NOTE.—The following rules are deduced immediately from the foregoing principles.

165. To find the Percentage when the Base and Rate are given.

RULE.

Multiply the base by the rate.

EXAMPLES.

1. What is 25% of 40 *lbs.?*   *Ans.* 40 *lbs.* × .25 = 10 *lbs.*
What is

2. 9% of 711 *lbs.?*	10. 3.7% of 140.5 ?
3. 3¼% of $1,200 ?	11. 125% of $14.40 ?
4. 7% of 810 *yds.?*	12. 210% of 97.4 *kilom.?*
5. 3¾% of 392 *miles?*	13. 93½% of 4.56 *lbs.?*
6. 6% of $500?	14. 62¼% of $185.57 ?
7. 12% of $1/12 ?	15. 15¾% of $136.64 ?
8. 11% of 31.25 *kilog.?*	16. 25% of £.37?
9. 14% of 875 *ft.?*	17. 18¼% of 84 *yds.?*

18. A man's income is $1700 a year, and he spends 75% of it; how many dollars does he spend?   *Ans.* $1275.

19. A bought 320 acres of land, and sold 62½% of it; how many acres did he sell?

20. What is 75% of £28 16s. 8d.?

166. **To find the Amount, or the Difference, when the Base and the Rate are given.**

RULE.

To find the amount, multiply the base by 1 plus the rate.   To find the difference, multiply the base by 1 minus the rate.

NOTE.—The *amount* can be formed by adding the percentage to the base, and the *difference* by subtracting the percentage from the base.

EXAMPLES.

1. What is the amount of 694 *lbs.* increased by 10% of 694 *lbs.?*   *Ans.* 694 *lbs.* × 1.10 = 763.4 *lbs.*

7

2. A paid $175 for a horse, and sold him at an advance of 20%; what was the selling price?

$$Ans.\ \$175 + 20\% \text{ of } \$175 = \$210.$$

3. If 464 *yds.* is diminished by 16% of itself, what is the difference?　　　*Ans.* 464 *yds.* × .84 = 389.76 *yds.*

4. A farmer had 72 tons of hay, and sold 16⅔% of it; how much had he left?

$$Ans.\ 72\ T. - 16\tfrac{2}{3}\% \text{ of } 72\ T. = 60\ T.$$

What is the amount and what is the difference of_

5. $200 and 9% of $200?

6. 770 *lbs.* and 25% of 770 *lbs.?*

7. 48 *yds.* and 33⅓% of 48 *yds.?*

8. 42 *kilom.* and 8% of 42 *kilom.?*

9. 75 *hhds.* and 66⅔% of 75 *hhds.?*

10. 36 *yrs.* and 12% of 36 *yrs.?*

11. 152 *ft.* and 7.5% of 152 *ft.?*

12. 67 *mi.* and 11% of 67 *mi.?*

13. 64 *T.* and 35½% of 64 *T.?*

14. $86 and 16% of $86?

15. 72 *bu.* and 9% of 72 *bu.?*　16. 56 *fr.* and 18% of 56 *fr.*

17. A man bought a watch for $115, and sold it at a loss of 20%; what was the selling price?

18. Of a farm containing 118 *A.*, 55% is arable, and the rest is woodland; how many acres of woodland?

19. If cloth cost $4.50 a yard, for how much must it be sold per yard to gain 25%?

20. A grocer bought sugar at 10 *cts.* a pound, and sold it at a loss of 15%; what was the selling price per pound?

167. To find the Rate when Base and Percentage are given.

R U L E .

Divide the percentage by the base.

E X A M P L E S .

1. The percentage is $90.24, and the base is $752; what is the rate? $Ans.$ $90.24 ÷ $752 = .12 = 12%.

2. A merchant bought a sloop for $6,250, and sold it again for $7,750; what did he gain per cent? Ans 24%.

EXPLANATION.—Here the percentage is $7750 − $6250 = $1500, and the base is $6250 ; hence, the percentage is $1500 ÷ $6250 = .24.

3. A merchant bought cloth at $5 a yard, and sold it at $4.50 a yard; how much did he lose per cent. ?

$Ans.$ 10%.

What per cent of
4. $50 is $3 ?
5. £16 is £7 ?
6. 100 *yds* is 1⅜ *yds.?*
7. £1 is 2*s.* 8*d.?*

8. 4 *mi.* is 140 *rds. ?*
9. 8.25 *fr.* is 37.96 *fr.?*
10. 407¼ *bu.* is 505.3 *bu.?*
11. 1,248 *yds.* is 2,080 *yds.?*

12. A grocer bought 1,140 *lbs.* of sugar at 9½ *cts.* a pound, and sold the whole for $129.96 ; what did he gain per cent. ?

13. A farmer raised 250 *bu.* of oats, and sold all but 75 *bu. ;* what per cent. of his crop did he sell?

14. A. bought 216 *yds.* of muslin at 11¾ *cts.* a yard, and sold the lot for $31.72½ ; how many per cent. did he gain ?

15. The property of a bankrupt is worth $6,102.95, and his debts amount to $8,225 ; what per cent. can he pay ?

16. In a journey of 1,664 *mi.*, A. travels 208 *mi.* by stage and the rest by rail; what per cent. does he travel by rail?

17. A man's salary is $4,200; of this he spends 22% for fuel and rent, 15% for clothing, and $1,218 for other purposes: what per cent. of his salary has he left?

18. Out of a cask containing 66½ *gals.* 26.6 *gals.* were drawn; what per cent. remained in the cask?

168. To find the Base when the Rate and Percentage are given.

RULE.

Divide the percentage by the rate.

Note.—If the rate and amount are given, divide the *amount* by *1 plus the rate;* if the rate and difference are given, divide the *difference* by *1 minus the rate.*

EXAMPLES.

1. In a school 77 pupils are present, which is 87½% of the whole number on the roll; how many are there on the roll?   *Ans.* 77 ÷ .875 = 88.

2. A., in selling goods for $3,840, clears 20% on their cost; what was the cost price?

Ans. $3,840 ÷ 1.20 = $3,200.

3. A man sold a watch for $220, which was 20% below its value; what was its value?   *Ans.* $220 ÷ .80 = $275.

4. A man spends $1,230 a year, which is 82% of his salary; what is his salary?

5. A merchant sold cloth at $5.25 a yard, which was an advance of 25% on the cost price; what was the cost price?

6. The population of a town is 18,558, which is 20% greater than it was 5 years ago; what was it then?

7. In a mixture of wine and water there are 12½ *gals.* of water, which is 18¾% of the whole; how many gallons in all?

8. A.'s salary is $2,925, which is 65% of B.'s; what is B.'s salary?

9. This year there are 216 students in an academy, which is 20% more than there were last year; how many were there then?

10. The breadth of a field is 36½ *rds.*, which is 27% less than its length; what is its length?

11. By increasing the width of a sheet 25% it was made 75 *ft.* wide; how wide was it before?

12. A farmer sold 35% of his sheep at $4¼ each, and received for them $297.50; how many sheep had he?

MISCELLANEOUS PROBLEMS IN PERCENTAGE.

1. What is the difference between 5½% of $800 and 6½% of $1,050?

2. A farmer raises 850 *bu.* of wheat: he sells 18% of it at $1.25 a bushel, 50% of it at $1½ a bushel, and the remainder at $1.75 a bushel; what does he get for it all?

3. What is the difference between £2,971 and 37½% of £2,971?

4. A drover sold 40 sheep for $248, which was 55% advance on their cost; what did they cost him apiece?

5. What number is that which, being diminished by 30% of itself, gives 385?

6. A market-woman has 600 eggs, and a second market-woman has 15% more; how many has she?

7. A young man spent $18,750, which was 37½% of his inheritance; how much did he inherit?

8. The population of a certain town in 1870 is 15,340,

which is an increase of 18 per cent. on its population in 1860; what was the population in 1860 ?

9. The distance between two towns in France is 20% more than 4 *kilometers* ; what is the distance ?

10. A man had 11 *hectol.* of wine, but he lost 3% of it by leakage; how much had he left ?

11. A drover sold cows and sheep for $9,180; he received for his sheep 70% of what he got for his cows; what did he get for the cows ?

12. A farmer raises wheat and corn; his wheat crop is worth $1,036, which is 40% more than the value of his corn crop: what is the value of the corn crop ?

13. From a cask of wine 37% was drawn off and 33.39 gallons remained; how many gallons did it contain ?

14. A. invests 35% of his capital in land and has $13,000 remaining; what is his capital ?

15. An army loses 27% of its number in battle and has 22,630 men remaining; how many did it contain ?

16. A man bought a house for $1,225.50, which in 3 years rose in value 147%; what was it then worth ?

17. A man had $5,420; he bought goods with 37½% of it, and then lent 25% of the balance to a friend: how much had he left ?

18. A general had an army of 10,816 men, of whom he lost in action 6¼%; how many did he lose ?

19. A merchant bought 15 pieces of cloth, each containing 31¼ yards, and found on examination that 50% was damaged; how much was good ?

20. A man bought 75 acres of land at $42⅔ an acre and sold it all for $3,577½ ; what per cent. did he gain ?

21. A man has a capital of $12,500 : he puts 15% of it in stocks, 33⅓% in land, and 25% in mortgages; how many dollars has he left?

22. Henry Adam bought a house for $7,520, spent $4,220 on it for repairs, and $75 for other expenses; he then sold it for $17,427.12½; how much did he gain per cent.?

REVIEW QUESTIONS.

(**158.**) Define per cent. (**159.**) What is the rate per cent.? (**160.**) What is percentage? (**161.**) What is the base? (**162.**) The amount? (**163.**) The difference? (**164.**) Repeat the 5 principles of percentage. (**165–168.**) Give the corresponding rules.

II. COMMISSION.

DEFINITIONS.

169. Commission is a percentage paid to an agent for transacting business.

170. An Agent is one that transacts business for another. If he buys and sells merchandise, he is called a Commission Merchant, or Factor; if he buys and sells stocks, exchange, real estate, and the like, he is called a Broker; if he collects debts, taxes, and the like, he is called a Collector.

171. A Consignment is a quantity of merchandise sent to an agent for sale. The party that sends the goods is the Consignor, and the agent that receives them is the Consignee.

172. An Account of Sales is an account rendered by the Consignee to the Consignor. The amount due the consignor, after deducting commission and other expenses, is called the **net proceeds.**

173. All problems in Commission are solved by the rules for percentage.

The **base** on which commission is reckoned is what the agent expends, or collects, on account of his principal; except in buying and selling stocks, and the like, where the *commission* or *brokerage*, as it is called, is usually *based* on the par value.

EXAMPLES.

1. A factor received a consignment of flour, which he sold for $3,750; what was his commission at $4\frac{1}{2}\%$?

EXPLANATION.—Here the *base* is $3,750 and the rate $4\frac{1}{2}\%$; hence, the percentage, or commission, is $3,750 × .045 = $168.75.

2. A cotton broker sells 70 bales of cotton for $80 per bale; what is his commission at 3%?

3. A real estate broker sells a house for $23,750, at a commission of $1\frac{1}{8}\%$; what must he pay his principal?

4. A drover sells cattle for $4,250, at a commission of 4%; what does he pay the owner of the cattle?

5. A consignee sold 300 *bbls.* of flour at $7 per barrel; what is his commission at $2\frac{1}{2}\%$?

6. An auctioneer sold a house and furniture for $26,750; what was his commission at $1\frac{1}{8}\%$?

7. A. sold 500 pieces of cloth at $30 a piece and paid the owner $14,700; what was the rate of commission?

8. A factor sold 500 pieces of muslin, each containing 21 *yds.*, for 23 cents a yard; what was his commission at $2\frac{1}{2}\%$?

9. A real estate broker bought a house for $21,300, and, by direction of his principal, sold it again at an advance

of 20% on the cost; what was his total commission at the rate of 1¼% both for buying and for selling ?

10. A commission merchant receives $3,825 to invest in flour on a commission of 2%; what is his commission ?

EXPLANATION.—Here the amount is $3,825 and the rate 2% ; hence, the base is $3,825÷1.02 = $3,750, which is the cost of the flour. The commission is $3,750×.02 = $75.

11. A merchant in New York sends $12,600 to his factor in Chicago to buy flour, agreeing to pay 5% of the cost for commission; how much flour does he receive, the market price being $12 a barrel?

12. A manufacturer invested $22,050 in cotton; the market price was 15 cts. a pound, the commission 2½%, the freight and cartage 1¼%, and the insurance 1¼%; how many pounds did he buy ?

13. A commission merchant sold 120 pieces of muslin, each containing 30¼ yds., at 22 cts. a yard, at a commission of 5½%; how much did he receive ?

14. A. sold 75 firkins of butter, each weighing 56 lbs., at 22½ cts. a pound; what was his commission at 5% ?

15. An auctioneer sold 150 hhds. of sugar, each weighing 1,150 lbs., at $7 a hundred, on a commission of 1¼%; what was his commission, and how much did he pay to the owner ?

16. A broker receives $3,500 to buy cotton at 8 cts. a pound, his commission being 1¼%; how much does he buy ?

17. A broker bought a house, charging 2½% commission ; his bill was $3,224.24; what did he pay for the house ?

18. A commission merchant sold goods for $7,500 at 3½% commission; what did he make ?

Find the net proceeds of the following *accounts of sales*:

(19.)

Sales on account of

S. A. BETTS, *Romford, Ct.*

1873.	BUYER.	DESCRIPTION.	$	cts.
Jan. 4.	S. T. D.	78 *bu.* oats, @ 65c. . .	50	70
" 15.	P. Q. R.	115 *bu.* corn, @ 93c. . .	106	95
Feb. 11.	R. S. V.	1320 *lbs.* butter, @ 31c. . .	409	20
March 5.	P. S. Q.	2560 *lbs.* cheese, @ 14c. . .	358	40
		Gross amount	925	25

Charges.

Freight and cartage $39.54
Storage. 5.25
Commission on $925.25 @ 4½% . . . 41.63½

			86	42½
		Net proceeds	$838	82¾

(20.)

Sales on account of

Jos. P. QUINN, *Roxbury, N. Y.*

1873.	BUYER.	DESCRIPTION.	$	cts.
June 13.	C. D.	1563 *lbs.* pork, @ 10c. . . .		
" "	D. L.	408 *lbs.* ham, @ 19c.		
July 5.	S. K.	829 *lbs.* lard, @ 14c.		
		Gross amount . . .		

Charges.

Freight and cartage. $5.42
Commission at 5% 17.49⅖

		Net proceeds	$326	96⅖

(**169.**) What is commission? (**170.**) What is an agent? A commission merchant, or factor? A broker? A collector? (**171.**) What is a consignment? A consignor? A consignee? (**172.**) What is an account of sales? Net proceeds? (**173.**) By what rules are problems in commission solved? What is the base?

III. INSURANCE.

DEFINITIONS.

174. Insnrance is a guarantee of indemnity in case of loss by fire, or other casualty.

175. A Policy of Insurance is a written contract, in which a company agrees to pay a certain sum in case of loss.

176. The **Premium** is a percentage paid to the company as a compensation for the risk assumed.

NOTE.—There are various kinds of insurance; as, **Fire Insurance, Marine Insurance, Life Insurance, Accident Insurance,** and the like. These differ from each other in the nature of the risk assumed and in the mode of determining the *rate per cent.* The *rate per cent.* and the *times of paying the premium* having been fixed, the method of proceeding is essentially the same in all.

177. All problems of insurance are solved by the rules for percentage.

The *base* on which the *premium* is reckoned is the sum named in the policy.

EXAMPLES.

1. What premium must be paid to insure a house for \$8,000 for 1 year at $\frac{5}{8}\%$?

EXPLANATION.—Here, the *base* is \$8,000 and the *rate* $\frac{5}{8}\%$; hence, the percentage or *premium* is \$8,000 × $.00\frac{5}{8}$ = \$50.

2. What is the premium for insuring a ship and cargo valued at $73,850 at the rate of $3\frac{1}{4}\%$?

3. A. insured his house for 1 year for $8,000 at the rate of $\frac{1}{2}\%$, and his furniture for $3,000 at the rate of $\frac{7}{8}\%$; what was the total premium?

4. A merchant insures his store for $12,000 at the rate of $\frac{3}{4}\%$, and his stock of goods for $15,000 at the rate of $1\frac{1}{4}\%$; what is the entire premium?

5. B. owns $\frac{3}{4}$ of a cargo of goods worth $25,000, and insures his interest at $2\frac{1}{2}\%$; what is the premium?

6. A vessel and cargo valued at $37,900 are insured at the rate of 3%; what is the premium?

7. A shipping merchant sends wheat valued at $1,200, from Chicago to New York, which he insures at the rate of $1\frac{3}{4}\%$; what premium does he pay?

8. A man insures his house for $10,000 at $\frac{3}{8}\%$, his barn for $1,800 at $\frac{7}{8}\%$, and his furniture for $3,000 at $1\frac{1}{4}\%$; what premium does he pay?

9. A cargo of goods valued at $48,000, insured at $1\frac{1}{2}\%$, is injured to the amount of 37% of its value; what must the company pay, over and above the premium?

10. A merchant pays $2,340 on a vessel and cargo, the rate being $4\frac{1}{2}\%$; for what sum is he insured?

EXPLANATION.—Here, the percentage is $2,340 and the rate $4\frac{1}{2}\%$; hence, the base, or the sum insured, is $2,340 ÷ .045 = $52,000.

11. A man pays $87.50 for the insurance of house at $\frac{7}{8}\%$, and $50 for the insurance of furniture at $1\frac{1}{4}\%$; if both are destroyed by fire, how much will he receive?   .

12. Shipped 5,000 *bbls.* of flour worth $10.50 a barrel, and paid for insurance $2,887.50; what was the rate?

EXPLANATION.—Here, the base is \$52,500 and the percentage \$2287.50; hence, the rate is \$2887.50 ÷ \$52,500 = .055 = $5\frac{1}{2}$%.

13. A person 20 years of age is required to pay 1.38% per annum to insure his life; what must he pay each year on a policy of \$5,000?

14. A man 40 years of age wishes to insure his life for 5 *yrs.* and finds that the annual rate is 1.86%; how much must he pay a year on a policy of \$12,500?

15. What does it cost to insure a cargo of goods worth \$50,000, at the rate of $1\frac{3}{4}$%?

16. What does it cost to insure a vessel for \$13,000, at $2\frac{1}{8}$%, and her cargo for \$18,268.50, at $1\frac{7}{8}$%?

17. What is the premium on $\frac{3}{4}$ of a vessel, worth \$27,500, at $3\frac{3}{8}$%, and $\frac{3}{8}$ of the cargo, worth \$126,875, at 2%?

18. A merchant shipped 400 *bbls.* of fish, worth \$3.50 per barrel; for what amount must he insure, at 5%, to cover the value of the fish and the cost of insurance?

EXPLANATION.—Here, the difference is \$1,400 and the rate 5%; hence, the base is \$1,400 ÷ .95 = \$1,473.68$\frac{2}{5}$.

19. I have goods worth \$37,560.75, which I insure for $\frac{4}{5}$ of their value, paying \$178.20; what is the rate?

Ans. $\frac{593}{1000}$%.

20. A merchant insured 450 pieces of silk, each piece worth \$35$\frac{1}{4}$, at $4\frac{1}{2}$%; what was the premium?

21. I insure my house for \$8,000 for 3 years at $\frac{1}{4}$% per annum; what is the total premium?

REVIEW QUESTIONS.

(174.) What is insurance? **(175.)** What is a policy of insurance? **(176.)** What is the premium? Mention some of the different kinds of insurance. **(177.)** By what rules are the problems of insurance solved?

IV. PROFIT AND LOSS.

DEFINITIONS.

178. **Profit** and **Loss** are commercial terms indicating *gain*, or *loss*, in business transactions.

If the *selling price* of any article is greater than the *cost price* there is a *profit;* if the selling price is less than the cost price there is a *loss*. Both *profit* and *loss* are usually reckoned as percentages on the cost price, as a *base*.

179. The problems in *profit* and *loss* are solved by the rules for percentage.

EXAMPLES.

1. A merchant sold goods that cost $2,350, at an advance of 20%; what was his profit?

EXPLANATION.—Here, the base is $2,350 and the rate 20%; hence, the percentage or profit is $2,350 × .20 = $470.

2. A person entered into a speculation in which he in-vested $7,000, and cleared 15%; what was his profit?

3. Bought a horse for $325, and sold him again at an advance of 18%; what was the profit?

4. A merchant buys cloth at $6 per yard, and sells it again so as to clear 25%; what does he ask per yard?

EXPLANATION.—Here, the base is $6 and the rate 25%; hence, the amount, or selling price, is $6 × 1.25 = $7.50.

5. A person begins business with a capital of $18,000, and in one year increases it 12%; what is his capital at the beginning of the second year?

6. Bought 1,280 *lbs.* of sugar, at 7½ *cts.* a pound, and sold it at an advance of 25%; what was the profit?

7. Bought drugs for £150, and sold them at a profit of 250%; how much was the profit?

8. A merchant had goods worth $3,750, of which 66⅔ per cent. were destroyed by fire; what was his loss?

9. Bought a house for $24,500, expended $1,500 in repairs, $4,000 for furniture, and $800 for taxes; what must the whole be sold for, to get an advance of `14%?

10. Bought a 1000 *T.* of coal, at $5 a ton, and sold the same at a profit of 11%; what was the entire gain?

11. Bought butter at 18 *cts.* a pound, and lost by the purchase 33⅓ per cent; what was the selling price?

12. Bought a ⅔ of a factory, and then sold ⅓ of my share for $15,000, making 50% on the cost; what was the value of the whole factory at the time of purchase?

13. Sold goods at a loss of 8% on their cost, and received $8,280; what did they cost?

14. Sold a horse for $364, gaining 12% on its cost; what did it cost?

15. If I sell a horse for $240, and lose 20%, what should I have sold him for to gain 10%?

16. A. hires a piece of ground for $120, and spends $625 for Durham calves, and $250 for incidental expenses; how much must he get per head for the calves to gain 20%?

17. A stationer sold quills at $3.75 a thousand, and cleared 25% on their cost; how many per cent. would he have cleared, if he had sold them at $4.50 a thousand?

18. A. bought a factory for $8,000, and stocked it at a cost of $13,500: the building was destroyed by fire, but 60% of the stock was saved; what was his loss?

19. A grocer bought 500 bags of coffee, each bag containing 49¼ *lbs.*, at 12 *cts.* a pound, and sold it at a profit of 16⅔%; for what did he sell it?

20. A dealer bought 375 *T*. 15 *cwt*. of wool, at $75 a ton, and sold it at a profit of 20%; what did he gain?

21. A grocer bought 410 *bu*. of potatoes, at 96 *cts*. a bushel, and sold them for $492; what per cent. did he make?

22. A merchant sold goods for $900, losing 10%; what did they cost him?

23. A dealer sold 20 *bbls*. of flour for $7½ per barrel, and made 25%; what did all the flour cost him?

24. A person laid in 25 tons of coal during the summer, and saved thereby 25% on the winter price, which was $7.20 per ton; what did his coal cost him?

25. A merchant bought 250 *bbls*. of flour, at $5.50 per *bbl*., and sold it for $1,875; what per cent. did he make?

26. A grocer sold coffee at 40 *cts*. a pound, and cleared 25%; how should he have sold it to clear 12½%?

27. A. sold an engine for $24,000, and lost 4%; how much would he have received if he had made 10%?

28. A man sold a pair of horses for $224 and gained 40% by the transaction; what was their cost?

29. If flour at $9 a barrel gives a profit of 20%, what does it cost a barrel?

30. Bought books for $420, and sold them for $357; how many per cent. did I lose?

31. A cask containing 35 *gals*. lost 5⅔ *gals*. by leakage; what was the loss per cent.?

32. Mr. Johnson sold a carriage for $338.40 and gained 20%; what did he pay for it?

33. Bought muslin at 10 *cts*. a yard, and sold it at 12¼ *cts*. a yard; what per cent. did I gain?

34. A flock of sheep increases from 88 to 110 in a year; what is the gain per cent. ?

35. A grocer sold potatoes for $16.10, gaining 15%; if he had sold them for $18.20, how much would he have made above the cost price ?

REVIEW QUESTIONS.

(178.) What are profit and loss? When is there a profit in business? When a loss? What is the base? (179.) By what rules do we solve problems in profit and loss?

V. TAXES.

DEFINITION.

180. **A Tax** is a sum of money levied on a community for the support of government, or for public improvements.

The United States government is supported by taxes on imported goods, by taxes on certain manufactures, by sales of public lands, stamps, and the like.

Money for the support of state, city, county, and other subordinate branches of government is generally raised by a tax on property. In some of the States, however, a small personal tax is laid on each male citizen over 21 years of age ; this is called a poll-tax, and each person so taxed is called a poll.

TAX ON PROPERTY AND POLLS.

181. **A Tax on Property** is a percentage based on the **Assessed Value** of the property; that is, on its value as determined by officers appointed for the purpose, called **Assessors.**

Property is of two kinds, *real* and *personal*. Real estate is fixed property, as houses and lands; personal property is movable property, as money, bonds, cattle, furniture, and the like. Both real and personal property are liable to taxation.

METHOD OF LAYING A TAX.

182. The assessors prepare a list of all the taxable property, specifying the amount belonging to each property-holder within the district to be taxed, and also a list of polls. The aggregate valuation of all the taxable property is the *base* of taxation. The entire sum to be raised on property divided by this base gives the **Rate of Taxation.** The *amount of taxable property* held by any individual, multiplied by the *rate*, gives the amount of his *tax on property ;* this, increased by his *poll tax*, gives his *total tax*.

EXAMPLES.

1. A town is to be taxed $23,200, on an assessed valuation of $2,900,000; what is A.'s tax on an assessed valuation of $14,275 ?

EXPLANATION.—Here there is no poll tax ; consequently the *rate* is $23,200 ÷ $2,900,000 = .008; hence, A.'s tax is $14,275 × .008 = $114.20. *Ans.*

2. A town is to be taxed $13,848; its assessed valuation is $1,452,000, and it contains 390 polls each liable to a tax of $2: now if A. is liable for 2 polls and an assessed valuation of $7,820, what is his tax ?

EXPLANATION.—Here the poll tax is $780, and this taken from $13,848 gives $13,068 to be raised on property. The rate is therefore $13,068 ÷ $1,452,000 = .009; hence, A.'s property tax is $7,820 × .009 = $70.38, and this increased by a poll tax of $4, gives $74.38. *Ans.*

3. In the same town, B. is liable for 2 polls and is assessed for $8,290; what is his tax ?

4. In a school district, a tax of $375 is levied for the support of schools; what is A.'s tax on a valuation of $8,000, the entire valuation of the district being $120,000?

(**180.**) What is a tax? What is a poll tax? What are polls? (**181.**) What is a tax on property? Describe real and personal property. (**182.**) What is the method of levying a tax on property alone? What is the method when polls are counted? How is the *rate* determined?

———————◆———————

VI. SIMPLE INTEREST.

DEFINITIONS.

183. **Interest** is a *percentage* paid for the use of money.

Interest is reckoned at a certain rate per cent. for each year.

184. The **Principal** is the sum on which interest is reckoned; the **Rate** is the per cent. per annum which is to be paid; and the **Amount** is the sum of the principal and interest for a given time.

The rate of interest fixed by law is called the **Legal Rate**; interest reckoned at a greater rate is called **Usury.**

In about half the States of the Union the legal rate is 6%, and in about half of the remaining ones it is 7%. In some States two rates are fixed, between which any rate is legal if agreed to by the parties.

In reckoning time, *a month* is usually regarded as $\frac{1}{12}$ *of a year,* without reference to the number of days it may contain, and *a day* is regarded as $\frac{1}{30}$ *of a month.*

185. **Simple Interest** is interest reckoned on the principal only. **Compound Interest** is interest reckoned on the principal and also on the accrued interest as it falls due.

NOTE.—The rules for interest are but special applications of the general rules of percentage.

MENTAL EXERCISES.

1. What is the interest on $15 for 1 year at 6% per annum ? What is the principal in this case ? The rate ? The amount ? What is the interest on $25 for 1 year at 8%? What is the principal here? The rate? The amount ?

2. What is the interest on $30 at 7% for 1 year? for 2 years ? for 4 years ? for 9 years? In the last case, what is the principal ? The rate ? The interest ? The amount? The time ?

3. What is the interest on $16 for 5 years at 6% ? For 3 years at 7% ? For 4 years at 10% ? On $20 for 3 years at 6% ? On $50 for 2 years at 9% ?

186. To find the Interest when we know the Principal, the Rate, and the Time in Years.

Let it be required to find the interest on $860 for 3 *yrs.* at 6%.

EXPLANATION.—The interest on $860 for 1 *yr.* is equal to $860 × .06, or to $51.60 ; for 3 *yrs.* it is 3 times as great, or $51.60 × 3 = $154.80.

In like manner all similar cases may be treated, hence the following

$$\begin{array}{r} \$860 \\ .06 \\ \hline \$51.60 \\ 3 \\ \hline \$154.80 \end{array}$$

R U L E.

Multiply the principal by the rate, and that result by the time in years.

NOTE.—In applying this and the following rules, let all decimals be carried to three places, adding 1 to the last figure when the first figure rejected is equal to or greater than 5.

EXAMPLES.

1. What is the interest on $560 for 4 years at the rate of 7% per annum ? *Ans.* $560 × .07 × 4 = $156.80.

2. What is the interest on $794 for $3\frac{1}{2}$ *yrs.* at 7% ?

3. Find the interest on $8,942 for $3\frac{1}{4}$ years at $8\frac{1}{2}$%.

4. Compute the interest on $8,720 for $1\frac{1}{2}$ years at 7%.

5. Find the interest on $712 for 3 years at 6%.

6. What is the interest on $329.50 for 2 years at 7% ?

7. What is the interest on $986.30 for 1 year at $6\frac{1}{2}$% ?

8. Find the interest on $12,600 for 1 year at $4\frac{3}{8}$%.

9. Find the interest on $112.75 for 4 *yrs.* at 10%.

10. Find the interest on $2,884.25 for 3 *yrs.* at 5%.

11. Find the interest on $1,750 for 2 *yrs.* at 6%.

12. Find the interest on $396.50 for 5 *yrs.* at 8%.

To find the **Amount,** *add the interest to the principal.*

13. To what will $1,400 amount in 2 years at the rate of $3\frac{1}{2}$% per annum?   *Ans.* $1,400 + $98 = $1,498.

14. Find the amount of $4,186.25 for 1 year at $5\frac{1}{2}$%.

15. What is the amount of £168 for 1 year at 7% ?

16. Find the amount of $1,001.75 for 1 *yr.* at $6\frac{5}{8}$%.

                              Ans. $1,068.116.

17. What does $450 amount to in $2\frac{1}{4}$ years at 6% ?

18. Find the amount of $3,875.20 for 5 years at $4\frac{3}{4}$%.

19. What does £2,000 amount to in 7 years at $8\frac{1}{3}$% ?

20. To what will $7,500 amount in 5 *yrs.* at $3\frac{1}{4}$% ?

21. Find the amount of $736 for 4 *yrs.* at 10%.

22. Find the amount of $1,490 for $3\frac{1}{2}$ *yrs.* at 12%.

23. Find the amount of $2,714 for 5 *yrs.* at 5%.

24. Find the amount of $10,863$\frac{1}{2}$ for 1 *yr.* at 8%.

25. Find the amount of $16,314 for 2 *yrs.* at 7%.

26. Find the interest on $9,000 for 2 *yrs.* at 6%.

27. Find the interest on $4,000 for 3 *yrs.* at 10%.

187. To find the Interest when we know the Principal, the Rate, and the Time in Months.

Let it be requred to find the interest on $480 for 9 months at the rate of 7% per annum.

OPERATION.

EXPLANATION.—The interest on $480 for 1 year is $33.60; hence, the interest on the same sum for 1 month is $\frac{1}{12}$ of $33.60, or $2.80; consequently, the interest for 9 months is 9 times as great, or $25.20.

In like manner we may treat all similar cases; hence, the following

$480
.07
12)$33.60
$2.80
9
Ans. $25.20

R U L E.

Multiply the principal by the rate and divide the product by 12; then multiply the quotient by the number of months.

NOTE.—The time may be reduced to decimals of a year, and then the last rule can be used; but in most cases, and especially when the rate is 6%, the rule here given is the simpler.

E X A M P L E S.

1. What is the interest on $815 for 11 _mos._ at the rate of 7% per annum?

Ans. $815 × .07 ÷ 12 × 11 = $52.296.

2. What is the amount of $1,375 for 4 _mos._ at the rate of 5% per annum?

Ans. $1,375 + $22.917 = $1,397.917.

3. Find the interest on $1,742.10 for 7 _mos._ at 5½%.

4. Find the interest on $840 for 9 _mos._ at 4½%.

5. Find the interest on $711 for 14 _mos._ at 8%.

6. Find the amount of $1,285 for 16 _mos._ at 6%.

7. Find the interest on $748 for 8 _mos._ at 6%.

Ans. $29.92.

8. Find the amount of $4,316 for 9 *mos.* at 10%.

9. Find the amount of $2,872 for 7 *mos.* at 5%.

10. Find the interest on $911.50 for 3 *mos.* at 8%.

11. Find the interest on $79.48 for 10 *mos.* at 4%.

12. Find the amount of $693.25 for 9 *mos.* at 7%.

13. Find the interest on $748 for 8 *mos.* at 6%.

SIMPLIFICATION.—When the rate per annum is 6%, the rate per month is $\frac{1}{2}$%, or, .005. In this case we multiply the principal by half the number of months and then divide by 100, or what is the same thing, we divide half the number of months by 100 and multiply the principal by the resulting quotient. Thus, in example 13, we have $748 × .04 = $29.92. *Ans.*

14. What is the interest of $890 for 10 months at 6% ?

15. Find the interest on $1,175 for 14 months at 6%.

16. Find the interest on $8,742.75 for 9 *mos.* at 6%.

17. Find the interest on $846 for 15 months at 6%.

18. What is the interest on $750 for 14$\frac{1}{2}$ months at 6% ?

19. Find the amount of $872 for 10 months at 6%.

20. Find the amount of $942 for 15 *mos.* at 6%.

21. Find the interest on $1,796 for 17 *mos.* at 6%.

NOTE.—If the time is given in *months* and *days*, or in *years*, *months* and *days*, reduce it to months and decimals of a month and proceed as before.

The operation of reduction can always be performed mentally.

22. What is the interest on $480 for 3 *yrs.* 4 *mos.* 21 *da.* at the rate of 7% ?

EXPLANATION.—Here we see that 3 *yrs.* 4 *mos.* is 40 *mos.*, and that 21 *da.* is $\frac{21}{30}$ or .7 of a month ; hence, the time in months is 40.7 *mos.* Finding the interest on $480 for 1 month, we have $2.80; hence, $2.80 × 40.7 = $113.96. *Ans.*

Find the interest

23. On $1,640 for 4 years, 5 months, and 12 days at 7%.

24. On $2,306 for 1 year, 7 months, 27 days, at 5%.

25. On $1,260 for 3 years and 6 days at 7%.

26. On $1,620 for 5 *yrs.* 24 *da.* at 4%.

27. On $675.89 for 3 *yrs.* 6 *mos.* 6 *da.* at 8%.

28. On $864.768 for 9 *mos.* 25 *da.* 6¼%.

29. On $100 for 1 *yr.* 3 *mos.* 10 *da.* at 6%.

30. On $1,000 for 9 *mos.* 15 *da.* at 7%.

31. On $1,700 for 2 *yrs.* 3 *mos.* 10 *da.* at 5%.

32. On £2,500 for 5 *mos.* 18 *da.* at 4½%.

33. On $450 for 3 *yrs.* 6 *mos.* 18 *da.* at 5%.

34. On $710 for 3 *yrs.* 10 *mos.* at 7%.

35. On $1,766 for 1 *yr.* 4 *mos.* 18 *da.* at 6%.

EXPLANATION.—Here the rate is 6% ; hence, we divide half the number of months by 100, which gives .083, and multiply the principal by this result. Thus, $1,766 × .083 = $146.578. *Ans.*

36. What is the interest on $14.50 for 19 days at 6%?

$$Ans. \quad \$14.50 \times .3167 = \$0.0459.$$

37. The amount of £10,000 for 3 *yrs.* 7 *mos.* 12 *da.* at 6%?

38. What is the interest of $2,300 from May 3d, 1870, to January 15th, 1873, at 6%?

39. What is the amount of $3,150 from August 16th, 1861, to May 1st, 1869, at 6%?

40. The amount of $5,675 for 3 *yrs.* 9 *mos.* 24 *da.* at 6% ?

41. The interest on $3,000 for 4 *yrs.* 8 *mos.* 6 *da.* at 6% ?

42. The interest on $2,500 for 1 *yr.* 8 *mos.* 12 *da.* at 7% ?

43. The amount of $3,500 for 2 *yrs.* 4 *mos.* 18 *da.* at 4½% ?

44. The amount of $850 for 1 *yr.* 9 *mos.* 15 *da.* at 7½% ?

45. The interest on $1,800 for 2 *yrs.* 3 *mos.* 10 *da.* at 8% ?

46. The interest on $2,100 for 9 *mos.* 12 *da.* at 4½% ?

47. The interest on $1,100 for 11 *mos.* 6 *da.* at 7% ?

48. The interest on $3,000 for 3 *yrs.* 6 *mos.* 18 *da.* at 5% ?

49. The amount of $4,000 for 1 *yr.* 3 *mos.* 12 *da.* at 4% ?

50. The amount of $5,000 for 1 *yr.* 1 *mo.* 6 *da.* at 7% ?

188. To find the Interest for Days at the Rate of 6%.

EXPLANATION.—In business transactions, interest for days is computed on the supposition that 30 days make 1 month, and 12 months 1 year, that is, that the year consists of 360 days. In this case, when the rate is 6%, the interest on $1 for 6 days is 1 *mill ;* hence, the following

RULE.

Multiply the principal by the number of days, divide the result by 6, and then move the decimal point three places to the left

EXAMPLES.

1. What is the interest on $84.60 for 15 *da.* at 6% ?

SOLUTION.—We multiply $84.60 by 15 and divide the result first by 6 and then by 1,000 ; hence, $86.40 × 15 ÷ 6,000 = $.212. *Ans.* The operation can often be simplified by cancellation.

2. What is the interest on $175.20 for 18 days at 6% ?

3. What is the amount of $144 for 25 days at 6% ?

4. What is the interest on $710 for 11 *da.* at 6% ?

5. On $334.56 for 13 *da.* at 6% ?

6. On $511.27 for 17 *da.* at 6% ?

7. On $3,942.75 for 18 *da.* at 6% ?

NOTE.—Having found the interest for days at the rate of 6%, we may find the interest at other rates by the method of aliquot parts. Thus, to find the interest at 5%, we diminish the interest at 6% by its *sixth* part ; to find the interest at 7%, we increase the interest at 6% by its sixth part. The interest at 4½% and at 7½% are respectively equal to the interest at 6%, diminished and increased by its fourth part. The interest at 4% and at 8% are respectively equal to the interest at 6%, diminished and increased by its third part, and so on.

8. Find the interest on $960 for 24 *da.* at 5% and at 7%.

SOLUTION.—The interest at 6% is $3.84 and $\frac{1}{6}$ of this is $.64; hence, at 5% the interest is $3.84 − $.64 = $3.20, and at 7% it is $3.84 + $.64 = $4.48.

9. Find the interest on $1,230 for 84 *da.* at 5% and at 7%.   *Ans.* $14.35 and $20.09.

10. On $960 for 66 days at 4½% and at 7½%.
11. On $648 for 54 *da.* at 4½% and at 7½%.
12. On $362.50 for 27 *da.* at 4% and at 8%.
13. On $187.75 for 90 *da.* at 3% and at 9%.
14. On $124.20 for 63 *da.* at 5% and at 7%.

189. To find Accurate Interest for Days.

EXPLANATION.—The preceding method gives a result too great by its $\frac{1}{73}$ part; to find the accurate interest, we may diminish the result found in accordance with the preceding rule by its $\frac{1}{73}$ part, or we may use the following

RULE.

Find the interest for 1 year at the given rate; then multiply the result by the number of days and divide the product by 365.

EXAMPLES.

1. What is the accurate interest on $803 for 35 days at the rate of 7% per annum?

SOLUTION.—The interest of $803 for 1 year at 7% is $56.21; hence, $56.21 × 35 ÷ 365 = $5.39. *Ans.*

2. Find the accurate interest on $584 for 70 *da.* at 7%.

3. On $876 for 105 *da.* at 8% per annum.

4. On $3,712.25 for 93 *da.* at 7%.

5. On $112.70 for 63 *da.* at 6%.

6. On $396.64 for 63 *da.* at 7%.

7. On $1,815 for 93 *da.* at 5%.

190. To find the Rate when we know the Principal, the Interest, and the Time in Years.

Let it be required to find the rate when the principal is $712, the interest $128.16, and the time 3 *yrs.*

EXPLANATION.—The interest on $712 for 3 *yrs.* at 1% is $21.36 ; but the given interest is $128.16, that is, it is 6 times as great ; hence, the required rate is six times 1%, or 6%.

OPERATION.

$$\$712 \times .01 \times 3 = \$21.36.$$

$$\$128.16 \div \$21.36 = 6\%.$$

In like manner all similar cases may be treated ; hence, the following

RULE.

Find the interest at 1% on the principal for the given time ; then divide the given interest by the result.

EXAMPLES.

What is the rate per annum when

1. The interest on $950 for 16 *mos.* is $88.66⅔ ?

2. The interest on $380 for 1 *yr.* 4 *mos.* is $22.80 ?

3. The interest on $8,726 for 1½ *yrs.* is $916.23 ?

4. The interest on $712 for 3 *yrs.* is $128.16 ?

5. The interest on $329.5 for 2 *yrs.* is $46.13 ?

6. The interest on 794 for 3½ *yrs.* is $194.53 ?

7. The interest on $450 for 3 *yrs.* 6 *mos.* 18 *da.* is $79.87½ ?

191. To find the Time when the Principal, the Rate, and the Interest are given.

Let it be required to find the time in which the interest on $1,200 at 6% will be equal to $120.

EXPLANATION.—The interest on $1,200 for 1 year is $72. But $72 is contained in the given interest $120, $1\frac{2}{3}$ times; consequently, the required time is $1\frac{2}{3}$ times 1 year; that is, it is 1 yr. 8 mo.

OPERATION.

$$\$1200 \times .06 = \$72.$$

$$\frac{\$120}{\$72} = 1\frac{2}{3}\,yrs. = 1\,yr.\ 8\,mos.$$

In like manner we may treat all similar cases; hence, the following

RULE.

Find the interest on the principal for 1 year at the given rate; then divide the given interest by the result.

EXAMPLES.

Find the time in which the interest

1. On $712 at 6% will be $128.16.
2. On $8,942 at $8\frac{1}{2}$% will be $2,470.22$\frac{1}{4}$.
3. On $329.50 at 7% will be $46.13.
4. On $980 at 6% will be $44.10.
5. On $1,175 at 6% will be $82.25.
6. On $846 at 6% will be $63.45.
7. On $872 at 6% will be $915.60.
8. On $1,500 at 7% will be $210.
9. On $3,000 at 5% will be $600.

192. To find the Principal, when the Interest, the Rate, and the Time are given.

Let it be required to find the principal that will give $65 interest in 20 *mos.* at 6% per annum.

EXPLANATION.—The interest on $1 for 20 *mos.* is 10 *cts.;* now if $1 draws 10 *cts.* in the given time, it will require $650 to draw $65 in the same time; hence, $\frac{\$65}{10\ cts.}$ is equal to the number of dollars in the required principal.

OPERATION.

$$\$1 \times .06 \times 1\tfrac{2}{3} = \$0.10$$

$$\frac{\$65}{\$0.1} = \$650.$$

$$\therefore \$650. \ Ans.$$

In like manner we may reason on all similar cases; hence the following

RULE.

Find the interest on $1 for the given time at the given rate; then divide the given interest by the result.

What is the principal on which the interest

  1. At 5% for 18 *mos.* is $157.50?

  2. At 6% for 2 *yrs.* 6 *mos.* is $450?

  3. At 4½% for 3 *yrs.* 4 *mos.* is $412.50?

  4. At 8% for 27 *mos.* is $324?

  5. At 7% for 20.4 *mos.* is $297.50?

  6. At 6% for 15⅓ *mos.* is $7.66⅔?

MISCELLANEOUS EXAMPLES.

1. The interest on a certain sum for 4 years, at 7 per cent., is $266; what is the principal?

2. The interest on $3,675, for 3 years, is $771.75; what is the rate?

3. The principal is $459, the interest $183.60, and the rate 8 per cent.; what is the time?

4. The interest on a certain sum for 3 years, at 6 per cent., is $40.50; what is the principal?

5. The principal is $918, the interest $269.28, and the rate 4 per cent.; what is the time?

6. What sum of money must be placed at interest at 7%, for 3 *yrs.* 9 *mos.*, that the interest may be $393.75 ?

7. In what time, at 7 per cent., will a mortgage of $8,000, whose interest is unpaid, amount to $9,120 ?

8. If I buy a house for $5,620 and receive $1,803 for rent in 2 *yrs.* 3 *mos.* 15 *da.*, what rate of interest do I get for my money ?

9. What sum of money, at 6%, will produce, in 2 *yrs.* 9 *mos.* 10 *da.*, the same interest that $350 produces, at 8%, in 3 *yrs.* 10 *mos.* 5 *da.?*

10. In what time will $5,000 at 7%, produce the same interest as $9,625 at 6½%, in 4 *yrs.* 5 *mos.* 18 *da.?*

ANNUAL INTEREST.

193. Annual Interest is simple interest on the principal and also on each year's interest from the time it falls due to the time of settlement.

This mode of computation is legal in some of the States when notes are made payable " with interest annually."

EXAMPLES.

1. What is the interest on a note for $600 at 6%, payable in 3 years *with interest annually?*

SOLUTION.—The interest on $600 for 3 *yrs.* is $108; the interest on $36 (*the first year's interest*) for 2 *yrs.* is $4.32 ; and the interest on $36 (*the second year's interest*) for 1 *yr.* is $2.16 : hence, the entire interest is equal to $108 + $4.32 + $2.16 = $114.48. *Ans.*

2. What is the interest on a note for $1,200 at 7%, payable in 4 *yrs.* *with annual interest ?*

3. What is the interest on a note for $980 at 8%, payable in 4 *yrs.* *with annual interest ?*

NOTES.

194. A **Promissory Note** is a written promise to pay a sum of money, either on demand, or at some specified time.

The person who signs the note is called the **Maker**, and the party that has legal possession of it is called the **Holder**.

195. A **Negotiable Note** is one that is payable either to **order**, or to **bearer**.

FORM OF A NEGOTIABLE NOTE.

$375.

New York, October 16, 1873.

For value received I promise to pay to John Doe, or order, three hundred and seventy-five dollars on demand, with interest at 7 per cent.

Richard Roe.

In this case John Doe, the person named in the note, is called the **Payee**; he can transfer it by writing his name across the back; he is then called an **Indorser**, and is obliged to pay the money when it falls due, if the maker of the note, Richard Roe, fails to do so.

196. The **Face** of a note, or bill, is the sum named in it. Thus, in the note just above described, the *face* is $375.

PARTIAL PAYMENTS.

197. A **Partial Payment** is a payment of a part of the amount due on a note, or other written obligation to pay money.

The date and the amount of each partial payment is *indorsed*, that is, *written on the back of the note, or obligation*, and is to be taken into account in making the settlement.

METHODS OF SETTLEMENT.

198. The following method of settling a note, or other interest-bearing obligation on which partial payments have been made, has been sanctioned by the Supreme Court of the United States, and is now adopted in New York, Massachusetts, and many other States:

SUPREME COURT RULE.

I. Find the amount of the given principal up to the time when the sum of the partial payments is equal to, or exceeds, the interest then due; from this result subtract the sum of the partial payments to the time considered.

II. Take the remainder for a new principal and proceed as before, continuing the operation to the time of final settlement.

EXAMPLES.

1. On a note dated May 1, 1866, for $1,200 at 6%, were the following indorsements: Nov. 1, 1866, $100; Mar. 1, 1867, $20; Sept. 1, 1868, $180: what was due on the note Nov. 1, 1869?

OPERATION.

Given principal	$1200
Int. to Nov. 1, 1866 (6 *mos.*)	$36
Amount	$1236
1st payment	$100
1st new principal	$1136
Int. to Sept. 1, 1868 (22 *mos.*)	$124.96
Amount	$1260.96
Sum of 2d and 3d payments	$200
Second new principal	$1060.96
Int. to Nov. 1st, 1869 (14 *mos.*)	$74.267
Amount due Nov. 1st, 1869	$1135.227

EXPLANATION.—We compute the interest on $1,200 from the date of the note to the time of the first payment, and because the first payment is greater than the interest then due, we add the interest to the principal and subtract the first payment, which gives $1,136 for a new principal. We then see, by inspection, that the interest on the new principal from Nov. 1, 1866, to Mar. 1, 1867 (4 *mos.*), is greater than the second payment; we therefore compute the interest on the new principal to the time of the third payment, add it to the principal, and from the result subtract the sum of the 2d and 3d payments, which gives another new principal. We then find the amount of this principal up to the time of settlement.

2. On a note dated May 1, 1866, for $350 at 6%, were the following indorsements:

> Dec. 25, 1866, received $50.
> Sept. 1, 1868, " $20.
> June 13, 1869, " $100.

What was due April 13, 1870?

3. On a note dated July 1, 1869, for $700 at 7%, were the following indorsements:

> July 1, 1870, received $200.
> July 1, 1871, " $400.

What is due July 1, 1872?

4. On a note dated May 1, 1860, for $2,000 at 7%, were the following indorsements:

> May 1, 1861, received $500.
> May 1, 1862, " $450.
> May 1, 1863, " $750.
> May 1, 1864, " $400.

What was due May 1, 1865?

Ans. $311.761.

8

5. On a note dated Aug. 1, 1870, for $1,000 at 5%, were the following indorsements:

> Dec. 1, 1870, received $310.25.
> April 1, 1871,   " $225.50.
> Aug. 1, 1872,   " $400.00.

What was due Jan. 1, 1873 ?

6. On a note dated Sept. 18, 1873, for $7,000 at 6%, were the indorsements: July 6, 1874, $500; Sept. 24, 1875, $1,500; and Dec. 6, 1875, $1,000: what was due July 12, 1876 ?

7. On a note dated Jan. 6, 1875, for $1,280 at 7%, were the indorsements: July 18, 1875, $175; Dec. 12, 1875, $375; and July 24, 1876, $400: what was due July 9, 1877 ?

8. On a note dated June 15, 1874, for $1,500 at 7%, were the indorsements: Dec. 15, 1874, $300; May 30, 1875, $300; and Dec. 18, 1875, $400: what was due July 15, 1877 ?

When partial payments are made on interest-bearing obligations due within a year, the balance is usually adjusted amongst business men by the following rule, called

MERCANTILE RULE.

Find the amount of the principal from the date of the note to the time of settlement ; find the amount of each payment from the time it was made to the time of settlement, and subtract their sum from the first result.

NOTE.—In applying this rule, the times are reduced to days, and the interest is then computed by the rule for days.

9. On a note dated Jan. 1, 1873, for \$1,000 at 7%, were the following indorsements:

> Feb. 15, 1873, received \$200.
> May 16, 1873, " \$400.

What was due Aug. 14, 1873 ?

SOLUTION.

Amt. of \$1000 for 225 *da.* \$1043.75
Amt. of \$200 for 180 *da.* \$207
Amt. of \$400 for 90 *da.* . . . \$407

Sum of amts. of payments \$614

Balance due Aug. 14, 1873 . . . \$429.75

10. On a note dated Jan. 1, 1873, for \$800 at 6%, are the following indorsements:

> Feb. 6, 1873, received \$200.
> April 30, 1873, " \$210.

What is due on the note June 5, 1873 ?

REVIEW QUESTIONS.

(**183.**) What is interest? (**184.**) What is the principal? The rate? The amount? What is the legal rate? What is usury? How is time reckoned? (**185.**) What is simple interest? Compound interest? (**186.**) Rule for interest when the principal, rate, and time in years are given? (**187.**) When the time is given in months? When in years, months, and days? (**188.**) Rule for interest when the time is given in days? (**189.**) Rule for accurate interest? (**190.**) Rule for finding the rate? (**191.**) Rule for finding the time? (**192.**) Rule for finding the principal? (**193.**) What is annual interest? (**194.**) What is a promissory note? The maker? The holder? (**195.**) What is a negotiable note? Payee? Indorser? (**196.**) What is the face of a note or other obligation? (**197.**) What is a partial payment? (**198.**) What is the Supreme Court rule? The Mercantile rule?

VII. COMPOUND INTEREST.

DEFINITIONS.

199. Compound Interest is interest computed on the principal and also on the accrued interest as it falls due.

Interest may be added to principal at the end of each *year, half year*, or other *fixed period*. Unless otherwise stated, the period is supposed to be a year.

200. From principles already explained we have the following

RULE FOR COMPOUND INTEREST.

I. Find the amount of the given principal for the first period ; then find the amount of this result for the second period ; and so on to the end of the given time ; the final result will be the total amount.

II. From the total amount subtract the given principal and the remainder will be the compound interest.

EXAMPLES.

1. What is the compound interest on $642 for 2 years at 6% per annum, interest being compounded annually ?

EXPLANATION.—Here the period is 1 year, and the amount of $1 for that period is $1.06 ; multiplying $642 by 1.06, we find the amount for the first period, $680.52 ; multiplying this by 1.06, we find its amount for the second period, $721.351. Subtracting the original principal, we find $79.351 for the required interest.

OPERATION.

$642 . . . . Principal,
1.06
————
$680.52 . . . . 1st amount,
1.06
————
$721.351 . . . . Total amount,
$642
————
$79.351 . . . Compound int.

2. What is the compound interest on $918 for 3 years at 6% per annum, interest being compounded annually?

3. What is the *amount* of $650 for 4 years at 6% per annum, interest being compounded semi-annually?

. EXPLANATION.—Here the period is one half of a year, and the amount of $1 for each period is $1.03. Proceeding as before, we find the amount.

NOTE.—The operation of computing the amount of any sum for a given time may be shortened by the use of the following

TABLE,

Showing the amount of $1 at compound interest, for any number of periods from 1 to 20.

PERIODS.	2%.	2½%.	3%.	3½%.	4%.	5%.	6%.	7%.
1	1.0200	1.0250	1.0800	1.0350	1.0400	1.0500	1.0600	1.0700
2	1.0404	1.0506	1.0609	1.0712	1.0816	1.1025	1.1236	1.1449
3	1.0612	1.0769	1.0927	1.1087	1.1249	1.1576	1.1910	1.2250
4	1.0824	1.1038	1.1255	1.1475	1.1699	1.2155	1.2625	1.3108
5	1.1041	1.1314	1.1593	1.1877	1.2167	1.2763	1.3382	1.4026
6	1.1262	1.1597	1.1941	1.2293	1.2653	1.3401	1.4185	1.5007
7	1.1487	1.1887	1.2299	1.2723	1.3159	1.4071	1.5036	1.6058
8	1.1717	1.2184	1.2668	1.3168	1.3686	1.4775	1.5938	1.7182
9	1.1951	1.2489	1.3048	1.3629	1.4233	1.5513	1 6895	1.8385
10	1.2190	1.2801	1.3439	1.4106	1.4802	1.6289	1.7908	1.9672
11	1.2434	1.3121	1.3842	1.4600	1.5395	1.7103	1.8983	2.1049
12	1.2682	1.3449	1.4258	1.5111	1.6010	1.7959	2.0122	2.2522
13	1.2936	1.3785	1.4685	1.5640	1.6651	1.8856	2.1329	2.4098
14	1.3195	1.4130	1.5126	1.6187	1.7317	1.9799	2.2609	2.5785
15	1.3459	1.4483	1.5580	1.6753	1.8009	2.0789	2.3966	2.7590
16	1.3728	1.4845	1.6047	1.7340	1.8730	2.1829	2.5404	2.9522
17	1.4002	1.5216	1.6528	1.7947	1.9479	2.2920	2.6928	3.1588
18	1.4282	1.5597	1.7024	1.8575	2.0258	2.4066	2.8543	3.3799
19	1.4568	1.5987	1.7535	1.9225	2.1068	2.5270	2.0256	3.6165
20	1.4859	1.6386	1.8061	1.9898	2.1911	2.6533	2.2071	3.8697

4. What is the amount of \$820 for 6 *yrs.* at 4% per annum, interest being compounded semi-annually?

EXPLANATION.—In this case there are 12 periods and the rate for each period is 2%. From the table, we find that \$1 at the given rate amounts to \$1.2682 in that time, and because \$820 amounts to 820 times as much as \$1, we have, \$1.2682 × 820 = \$1,039.924. *Ans.*

5. What is the amount of \$900 for 9 years at 7% per annum, interest being compounded semi-annually?

6. What is the compound interest on \$1,850 for 3 *yrs.* at 8%, interest being compounded quarterly?

7. Find the amount of \$800 for 14 years at 7% per annum, interest being compounded annually.

Ans. \$2,062.80.

NOTE.—If the last period is fractional, compute the amount to the end of the next preceding period, and then find the amount of that result for the fractional period.

8. What is the amount of \$500 for 3 years 2 months, at 6% per annum, interest being compounded annually?

EXPLANATION.—The amount for 3 years is \$1.191 × 500 or \$595.50, and this in two months amounts to \$601.455. *Ans.*

9. What is the amount of \$1,200 for 4 *yrs.* 8 *mos.* at 7%, interest being compounded annually?

Ans. \$1,646.365.

10. What is the amount of \$1,350 for 5 *yrs.* 4 *mos.* at 6%, interest being compounded semi-annually?

Ans. \$1,850.55.

NOTE.—In using the table we retain the 4 decimal places given.

REVIEW QUESTIONS.

(**199.**) What is compound interest? How often may interest be added to principal? (**200.**) What is the rule for compound interest? Explain the use of the table.

VIII. DISCOUNT.

201. **Discount** is a percentage deducted from the face of a bill, debt, or note.

COMMERCIAL DISCOUNT.

202. **Commercial Discount** is a percentage deducted from the face of a bill of merchandise.

The face of the bill is the **Base** and the difference between this and the discount is called the **Net Proceeds.**

From definitions and preceding principles we have the following

RULE.

I. Multiply the face of the bill by the rate per cent. and the product will be the discount.

II. Subtract the discount from the face of the bill and the difference will be the net proceeds.

NOTE.—The net proceeds may be found by multiplying the face of the bill by 1 minus the rate per cent.

EXAMPLES.

1. What is the discount on a bill of $350 at 5%, and what is the net proceeds ?

   Ans. Dis. = $17.50 ; net proceeds = $332.50.

2. Sold a bill of merchandise amounting to $1,173, deducting 10% for cash; what was the net proceeds ?

3. Flour is sold on credit at $12.50 per barrel; what is the cash price, the discount being 15% ?

4. Find the discount on a bill of goods whose face is $1,200, at the rate of $2\frac{1}{2}\%$.

5. Sold a lot of goods amounting to $918, deducting 12½% for cash; what was the net proceeds?

6. What is the net proceeds of 56 tubs of butter, each weighing 42¼ *lbs.* at 22 *cts.* a pound, 5% off for cash?

7. Sold 50 *bbls.* flour at $7.50 per barrel, deducting 7½% for cash; what was the net proceeds?

8. Sold coal at $5.50 per ton, 10% off for cash; what was the cash price?

9. Sold 500 *bu.* of oats at 62½ *cts.* a bushel, 5% off for cash; what was the cash price?

PRESENT VALUE AND TRUE DISCOUNT.

203. The **Present Value** of a debt payable at a future time is a sum which, being placed at interest, will give an amount equal to the debt when it falls due. Thus, $100 is the present value of $107 due 1 year hence, interest being reckoned at 7%.

If the debt does not bear interest, its *amount* is the same as its *face;* if it bears interest, its amount is equal to its *face* together with the *interest up to the time it is due.*

The **True Discount** is the difference between the amount of the debt and its present value.

The method of finding the present value of a debt due at a future time is the same as finding the principal, when the amount, the rate, and the time are given; hence, the following

R U L E.

I. Divide the amount of the debt by 1 plus the product of the rate by the time in years; the quotient will be the present value.

II. Subtract the present value from the amount of the debt; the remainder will be the true discount.

EXAMPLES.

1. What is the present value of $1,500, due 1 *yr.* 4 *mos.* hence, money being worth 6% per annum ?

SOLUTION.—The rate, .06, multiplied by the time in years, 1⅓, equals .08; hence, $1,500 ÷ 1.08 = $1,388.889. *Ans.*

2. What is the true discount on $1,200, due 2 years hence, money being worth 7% ?

Ans. $1,200 — $1,200 ÷ 1.14 = $147.37,

3. What is the present value of a debt of $1,760, due 3 *yrs.* 6 *mos.* hence, at the rate of 6% per annum ?

4. Find the true discount on a debt of $1,141.25, due 7 *mos.* 15 *days* hence, at 6% per annum.

5. What is the true discount on $730, due in 2 years, at 5% per annum ?

6. Find the present value of a debt of $986, due 2 *yrs.* 8 *mos.* hence, at 6% per annum.

7. What is the true discount in the last example ?

8. What is the present value of $1,200, due in 1 *yr.* 4 *mos.*, at 7½% per annum ?

9. A debt of $1,400 is due in 9 *mos.*; what is the true discount, interest being computed at the rate of 6% ?

10. Find the present value of a note for $750, due in 1 *yr.* 8 *mos.* 12 *da.*, at 6%.

11. What is the present value of a note for $1,300, due in 2 *yrs.* 8 *mos.* at 7% ?

12. What is the present value of $10,000, due in 4 *mos.* 18 *da.*, at 4½% ?

13. What is the present value of $1,828.75, due in 1 year, and bearing 4½% interest?

14. What is the present value of a note for $4,800, due 4 *yrs.* hence, at 5% interest?

15. A. owes B. $3,456, payable Oct. 27, 1877; what ought A. to pay Aug. 24, 1877, interest being 6% per annum?

BANKS AND BANK DISCOUNT.

204. A **Bank** is an incorporated institution authorized by law to deal in money.

Some of the operations of banking are, receiving money for safe keeping, discounting notes and other evidences of indebtedness, issuing bills to circulate as money, buying and selling bills of exchange, gold and silver bullion, coin, and the like. Most banks are engaged in only a part of these operations.

205. Bank Discount is a percentage charged for advancing money on a note or other obligation payable at a future time; it is simply *interest in advance* on what the note or obligation will yieeld when it becomes *legally* due.

This operation of advancing money on notes or obligations not yet due is called **Discounting**.

206. A note is said to **Mature** when it becomes legally due, which is *three days* after it is nominally due. The three days that elapse after a note is *nominally* due before it is *legally* due are called **Days of Grace**.

The time at which a note falls due may be denoted by a double date. Thus, the expression, "Due July $7/10$," written at the foot of a note shows that it is *nominally* due on the 7th of July, and that it is *legally* due on the 10th of July.

NOTE.—If the last day of grace falls on Sunday, or on a legal holiday, the note is legally due in some of the States on the *preceding* day, and in others on the *following* day.

METHOD OF DISCOUNTING A NOTE.

207. The ordinary method of discounting a note is illustrated in the following example:·

Form of Note.

$600.

NEW YORK, *July* 7, 1877.

Sixty days after date I promise to pay to the order of Lucius Slocum six hundred dollars at the Fourth National Bank. Value received.

ROBERT BULL.

Due Sept. ⁵/₈, 1877.

EXPLANATION.—This note is supposed to be discounted on the day of its date. The holder, LUCIUS SLOCUM, endorses it by writing his name across its back, and delivers it to the proper bank officer. Interest is then computed on the face of the note, $600, for 63 days, the days of grace being included, and at the legal rate, which in New York is 7%. This sum, $7.35, is the *discount;* subtracting the discount from $600, we have $592.65, which is called the **Proceeds,** and this amount is paid over to LUCIUS SLOCUM. If the note is not paid before the close of the last day of grace, Sept. 8, a written notice, called a **Protest,** is sent to LUCIUS SLOCUM, and he then becomes liable for its payment.

EXAMPLE.

1. A note for $1,400, payable 60 days after date, is discounted at the rate of 7%; what is the proceeds?

EXPLANATION.—The interest of $1,400 for 63 days at 7% is $17.15; this is the discount: subtracting it from $1,400, we have $1,382.85, which is the *proceeds.*

NOTE.—If a note is not discounted on the day of its date, the discount is reckoned from the time of discount to the time of maturity.

EXAMPLE.

2. A note for $1,200, dated Sept. 5, 1877, and payable 90 days after date at 7%, is discounted Oct. 2, 1877; what is the proceeds? *Ans.* $1,184.60.

EXPLANATION.—Here the note is due Dec. $^4/_7$, and from Oct. 2 to Dec. 7 is 66 days   The interest on $1,200 for 66 days at 7% is $15.40 ; hence, the proceeds equals $1,200 — $15.40, or, $1,184.60.

From what precedes we have the following

RULE.

I. Compute interest on the face of the note, from the time of discount to the time of maturity; this will be the discount.

II. Subtract the discount from the face of the note; the result will be the proceeds.

NOTE.—If an interest-bearing note is discounted, the amount of the note at maturity is made the base on which discount is reckoned.

EXAMPLES.

3. What is the bank discount on a note for $670, payable 60 days after date at 6% ?

4. What is the bank discount on a note for $350, payable 90 days after date at 7% ?

5. What is the proceeds of a note of $1,000, payable at bank, 60 days after date, at 6% ?

6. A. has a note against B. for $1,728, payable 90 days after date, which he gets discounted at the rate of 7%; what does he receive ?

7. A note for $1,620, dated July 7, 1877, and payable 90 days after date, is discounted July 25, 1877, at 8%; what is the proceeds ?

NOTE.—In what precedes, interest has been computed on the supposition that 360 days make a year.   If accurate interest is required, as it is in many of the States, it may be found by diminishing the interest, computed as above, by its $\frac{1}{73}$ part, or, better still, by means of tables constructed for the purpose.

208. To find the Face of a Note payable at a Future Time, whose Proceeds shall equal a Given Sum.

To find the face of a note, payable 90 days after date, that will yield $500, the rate being 6%.

EXPLANATION.—The proceeds of $1 for 93 days at 6% is $.9845. But the entire proceeds must be $500; hence, the face of the note must be as many dollars as .9845 is contained in 500, that is, it must be $500 ÷ .9845, or $507.872.

In like manner all similar cases may be treated; hence, the following

RULE.

Divide the given sum by the proceeds of $1 for the given time and rate; the quotient will be the number of dollars required.

EXAMPLES.

1. Find the face of a note payable in 90 days at 7%, so that the proceeds shall be $2,050. *Ans.* $2,087.747.

2. Find the face of a note payable 60 days after date at 6% that will yield $500.

3. Find the face of a note payable 90 days after date at 8% that will yield $750.

4. Find the face of a note payable 30 days after date at 10% that will yield $1,000.

REVIEW QUESTIONS.

(**201.**) What is discount? (**202.**) What is commercial discount? Rule? (**203.**) What is the present value of a debt? What is true discount? Rule for present value and true discount? (**204.**) What is a bank? What operations may be carried on by a bank? (**205.**) What is bank discount? What is discounting? (**206.**) When does a note mature? What are days of grace? (**207.**) Explain the process of discounting a note. When and how is a note protested? Give rule for finding the bank discount and proceeds of a note. (**208.**) What is the rule for finding the face of a note that will yield a given sum?

IX. STOCKS AND BONDS.

DEFINITIONS.

209. A **Corporation** is an association of persons authorized, under certain restrictions, to transact business as an individual.

210. The **Capital Stock** is the money used in carrying on the business of a corporation. It is divided into equal parts called **Shares**.

The owners of the shares are called Stockholders.

211. A **Certificate of Stock** is a certificate signed by proper authority, showing that the party therein named owns a certain number of shares of the capital stock.

212. The **Par Value** of a stock is the value named on the face of the certificate. This is usually $100 per share, but it may be either more or less.

213. The **Market Value** of a stock is the price it will bring in open market.

If the market value is greater than the par value, the stock is said to be **at a Premium,** or **Above Par;** if the market value is less than the par value, it is said to be **at a Discount,** or **Below Par.**

214. **Dividends** are percentages on the capital stock paid to *stockholders* as *profits* on the business. **Assessments** are percentages that *stockholders* are called on to pay to meet losses and extraordinary expenses.

215. A **Bond** is a properly authenticated obligation to pay a sum of money at or before a certain time, with interest at fixed periods.

There are two classes of bonds: 1'. Bonds for whose payment the public faith is pledged, as National and State bonds; and 2°. Bonds for whose payment the property of some incorporated company is pledged, as Railroad bonds.

216. The face of a bond is usually $1,000, but it may be any sum, either greater or less.

Stocks and bonds are bought and sold in open market. The reported price, or **Quotation**, is the number of *per cent.* that the selling price is of the par value. Thus, if a stock or bond is quoted at $87\frac{1}{2}$, its selling price is $87\frac{1}{3}\%$ of its par value.

UNITED STATES BONDS.

217. The interest-bearing debt of the United States is represented by bonds issued at separate times, payable at different dates, and bearing various rates of interest. Some bonds carry certificates of interest, called **Coupons**, and pass from hand to hand like bank bills; others are **Registered** on the books of the treasury, and can only be transfered by a change of record.

United States Bonds are designated by giving their date of payment, and, if necessary, their interest. Thus, the term "U. S. 6's, 81 R." stands for "Registered 6% Bonds of the United States, payable in 1881;" the term "U. S. 5-20's, 65 C." stands for "United States Coupon Bonds, payable at the option of the government at any time between 5 and 20 years after their date," that is, at any time between 1870 and 1885; the term "U. S. 10–40's" stands for "United States Bonds, payable at any time between 10 and 40 years after their date," which is 1864. All the 5–20's bear 6% interest, and all the 10–40's bear 5% interest, payable in gold.

218. All problems relating to the buying and selling of stock and bonds are solved by the rules for percentage and interest.

NOTE.—In what follows the par value of each share of stock is supposed to be $100, and the par value of each bond $1000; the brokerage, which is paid by the party for whom the purchase or sale is made, is supposed to be $\frac{1}{4}\%$, and is always computed on the par value of the stock or bond.

EXAMPLES.

1°. Applications of the Rules for Percentage (Arts. 165–168).

1. What is the market value of 120 shares of bank stock @ $53\frac{3}{4}\%$? *Ans.* $12,000 × .5375 = $6,450.

2. What is the cost, including brokerage, of 44 shares of bank stock @ $129\frac{1}{2}\%$? *Ans.* $4,400 × 1.2975 = $5,709.

3. Sold $5,000 in gold @ $112\frac{1}{4}\%$; what was the net proceeds after deducting brokerage. *Ans.* $5,600.

4. A. sold 160 shares railroad stock @ $92\frac{3}{4}\%$, and purchased with the proceeds bank stock @ $73\frac{3}{4}\%$, paying brokerage on both sale and purchase; how many shares did he receive? *Ans.* 160 × $92\frac{1}{2}$ ÷ 74 = 200.

5. What is the premium on 88 shares of bank stock @ 114%? *Ans.* $8,800 × .14 = $1,232.

6. What is the discount on 190 shares of railroad stock which is sold @ 89% ?

7. A. sold 93 bonds, U. S. 5–20's, R. @ $113\frac{1}{2}\%$, paying brokerage on the sale; what did he receive ?

8. A speculator buys 225 shares of Erie stock @ $30\frac{1}{4}\%$ and sells it again @ $31\frac{1}{2}\%$, paying brokerage on both transactions; what does he gain ?

9. If 170 railroad bonds cost (including brokerage) $192,100, what is their market rate ?

10. What is the cost, including brokerage, of $9,000 U. S. 6's 81 C. @ $112\frac{1}{4}$?

11. Sold $42,000 U. S. 5–20's C. for $46,830 net, after paying brokerage; what was the market price ?

12. How much gold @ 111⅝ can be bought for $8,930 in currency, brokerage not considered?

13. A broker sells 30 shares of bank stock @ 96½%, and 120 shares of railroad stock at 105%, retaining the brokerage; how much does he pay to his principal?

Ans. $15,457.50.

2°. *Applications of the rules for interest.*

14. If I buy a 6% stock @ 90%, what rate of interest do I receive on the investment?

EXPLANATION.—Here the *principal*, **$90**, yields an interest of **$6** per annum; hence, we have,

$$\$6 \div \$90 = 6\tfrac{2}{3}\%. \; Ans.$$

15. The market rate of a 5% stock is 85½%; if the purchaser pays brokerage, what rate of interest does he receive on his investment?

16. An 8% stock sells @ 112%; what rate of interest does it yield to the purchaser?

17. At what rate must an 8% stock be purchased to yield the purchaser 7% interest?

In this case the price of one share is a principal which, at the rate of 7%, yields $8.00 in one year; hence, (Prin. 5°, Art. **164**), we have,

$$\$8.00 \div \$.07 = 114\tfrac{2}{7}\%. \; Ans.$$

18. I wish to purchase a 5% stock on such terms that it will give me 7% on my investment; how much can I pay for the stock, including brokerage?

19. If I buy U. S. 6's, '81, at 112½, including brokerage, and sell the gold interest at 107¾% for currency, what rate of interest do I get on my investment?

Ans. $6 × 1.07¾ ÷ 112½ = 5.75%, *nearly.*

20. At what rate must I buy a 6% stock to get the same rate of interest as from a 5% stock at 75% ?   *Ans.* 90%.

REVIEW QUESTIONS.

(**209.**) What is a corporation? (**210.**) What is the capital stock? Shares? Stockholders? (**211.**) What is a certificate of stock? (**212.**) What is par value of a stock? (**213.**) What is the market value? When is a stock at a premium? At a discount? (**214.**) What are dividends? Assessments? (**215.**) What is a bond? Explain the two classes. (**216.**) How are bonds and stocks quoted? (**217.**) Give an account of U. S. bonds. What are coupons and coupon bonds? Registered bonds? (**218.**) By what rules do we solve problems in stock operations?

———◆———

X. EXCHANGE.

DEFINITIONS.

219. Exchange is a method of making payments at distant places by means of drafts, or bills of exchange.

The theory of settling accounts by exchange may be illustrated by the following simple case : A flour merchant of Chicago forwards $5,000 worth of flour to a shipper in New York, and at the same time a dry goods merchant of Chicago buys $5,000 worth of merchandise from a New York importer. The flour merchant draws his draft for $5,000 on the shipper and sells it to the dry goods merchant ; the latter forwards it to the importer, who presents it to the shipper and receives the money called for. In this way the debts in both cities are liquidated without the necessity of sending any money from either.

Dealers in exchange usually buy drafts on distant places and send them forward as a basis of credit; they then sell their own drafts drawn against this credit in sums to suit their customers.

220. A **Draft** or **Bill of Exchange** is a written order from one party to another to pay to a third party a certain sum of money at a specified time.

A Sight Draft is one that is payable on presentation ; a **Time Draft** is one that is payable at a specified time, or at a certain time after presentation.

NOTE.—In the latter case *three days grace* are allowed, but not in the former.

221. The **Drawer** or **Maker** is the party that draws the bill; the **Drawee** is the one on whom it is drawn; and the **Payee** is the party to whom the money is to be paid.

A party that buys a bill of exchange is called a **Buyer**, or **Remitter** ; if the payee, or any other party, writes his name across the back of the bill, he is called an **Indorser** ; and the party that has legal possession of the bill at the time of payment is called the **Holder.**

222. An **Acceptance** is an agreement on the part of the *drawee* to pay the draft at maturity. If he agrees to pay it, he writes the word **Accepted** across its face and *signs his name ;* he is then called an **Acceptor.**

223. An **Inland** or **Domestic Bill** is one in which both *drawer* and *drawee* reside in the same country.

A **Foreign Bill** is one in which the drawer and drawee reside in different countries.

224. The **Par of Exchange** between two places is the relative value of the principal units of currency of the two places.

Thus, £1 **Sterling** is equal in value to $4.8665, and this is the par of exchange between London and New York ; the **Franc** is equal to $0.193, and this is the par of exchange between Paris and New York ; the **Mark** is equal to $0.238, and this is the par of exchange between Berlin and New York.

NOTE.—If $4.8665 in New York will just purchase a bill of £1 on London, exchange on London is at par ; if it will buy a larger bill, exchange is against London and in favor of New York ; if it requires more than $4.8665 to buy a bill of £1, exchange is in favor of London and against New York.

225. The **Course of Exchange** is the variation in price of bills of exchange.

These variations are shown in the daily quotations published in the papers. Thus, on the 10th of August sterling exchange was quoted at $4.86, *sight;* that is, a sight bill on London was worth $4.86 for each pound sterling of its face.

INLAND OR DOMESTIC EXCHANGE.

226. **Inland** or **Domestic Exchange** is the method of making payments at distant places in the same country by means of *drafts* or *inland bills of exchange.*

Form of a Draft.

$500.

TROY, N. Y., *Aug.* 10, 1877.

At sight, pay to the order of William Whitman Five Hundred Dollars, value received, and charge the same to the account of

JOHN S. WESLEY.

To the Fourth National Bank,
      New York City.

NOTE.—To change the above to a *time draft*, say for 10 days, the words " *at sight* " are replaced by " *at ten days sight.*"

227. All problems relating to inland exchange can be solved by the rules for percentage and bank discount.

In applying the rules for percentage, the **Face of the Bill** is the *base*, the **Premium** or **Discount** is the *rate per cent.*, and the **Cost of the Bill** is the *amount*.

EXAMPLES.

1. What is the cost of a sight draft when exchange is $1\frac{1}{2}\%$ premium ?

SOLUTION.—$500 + $1\frac{1}{2}\%$ of $500 = $507.50. *Ans.*

2. What is the cost of a sight draft on New York for $1,250, exchange being 1¼% discount? *Ans.* $1,234.37½.

3. What is the market price of a sight draft on New York for $890, exchange being worth 101¼%?

EXPLANATION.—Here the rate of premium is 1¼%, and 1 plus the rate, that is, the amount per cent. is 1.0125; hence, the draft is worth $890 × 1.0125 = $901.12½. *Ans.*

4. Find the market value of a sight draft on New York for $1,800, exchange being 99%. *Ans.* $1,782.

NOTE.—Both exchange and bank discount are computed on the face of a time draft. They may be computed separately, or together, as is more convenient.

5. Find the cost of a draft on New York for $1,400, payable 60 days after sight, exchange being worth 102½%, and interest being reckoned at 7%.

OPERATION.

Amount of $1400 at 2½% premium . . . $1435.00
Interest on $1400 for 63 days at 7% . . . 17.15
Difference . . . . $1417.85 *Ans.*

6. What is the cost of a draft on New York for $2,400, payable 90 days after sight, interest being 10% and exchange 103%? *Ans.* $2,410.

7. Find the value of a draft on New York for $1,650, payable 60 days after sight, exchange being 98½% and interest 6%. *Ans.* $1,650 × .9745 = $1,607.925.

8. If exchange is 101½%, how large a sight draft can be bought for $7,900? *Ans.* $7,900 ÷ 1.01½ = $7,783.251.

9. What is the face of a sight bill that can be bought for $5,000, when exchange is 98½%?

SOLUTION.—$5,000 ÷ .985 = $5,076.14½. *Ans.*

10. What is the face of a draft at 60 days sight which costs $1,000, exchange being 103% and interest 6%?

EXPLANATION.—The interest on $1 for 63 days is $0.0105, and this taken from $1.03 gives $1.0195 as the cost of $1 of exchange. Hence, the face of the bill is $1,000 ÷ 1.0195 = 980.87\frac{3}{10}$. *Ans.*

11. What is the face of a draft at 30 days, which costs $2,000, exchange being 102% and interest 6%?

FOREIGN EXCHANGE.

228. Foreign Exchange is the method of making payments in foreign places by means of bills of exchange.

Form of Set of Exchange.

£500.   NEW YORK, *Aug.* 11, 1877.

At sight of this *first* of exchange (*second* and *third* of same date and tenor unpaid), pay to the order of Salmon Stoddard Five Hundred Pounds Sterling, value received, and charge the same to

JOHNSON LUMMIS.

To, Copely & Brothers, London.

NOTE.—Three copies constitute a *set*, each of which is forwarded by a different mail. The other copies differ from the one above given, in no respect, except that in one the words *first* and *second* change places, and in the other the words *first* and *third* change places.

229. Exchange betweeen the United States and foreign countries is computed by the method of *equivalents*. The equivalents are made known, both for sight and time bills, by the current or daily quotations. But as these equivalents are in gold, they must be converted into currency by the ordinary rules for percentage.

EXAMPLES.

1. What is the cost of a sight bill for £87, when £1 is worth $4.82 in gold, gold being worth 106% currency?

EXPLANATION.—Because £1 is worth $4.82 in gold, £87 is worth $4.82 × 87 = $419.34, and this is converted into currency by multiplying it by 1.06 ; hence, $419.34 × 1.06 = $444.50. *Ans.*

2. What is the cost of a sight bill on London for £750, when £1 is worth $4.85, gold being quoted at 108%?

3. What is the cost of a sight bill on London for £300, £1 being worth $4.86, gold 107% currency?

4. Find the cost of a 60 day bill on London for £315, when £1 at 60 days is worth $4.80, gold being quoted at 105?

5. How large a bill on London can be bought for $2,000 in currency, when sterling exchange is quoted at $4.85 and gold at 106?

EXPLANATION.—Here we convert $4.85 in gold into currency, which gives $4.85 × 1.06 = $5.14. Then, because it takes $5.14 in currency to buy £1 of exchange, we have $2,000 ÷ 5.14 = £389 2s. 1d. *Ans.*

6. How large a 60 day bill on London can be bought for $3,000, when 60 day bills are quoted at $4.84, and gold at 110?

7. What is the cost of a bill on Paris for 3,000 *francs*, when exchange is at the rate of $1 to 5.15 *francs*, and gold is worth 110?

Ans. ($5,000 ÷ 5.15) × 1.10 = $1,067,96$\frac{1}{10}$.

8. What is the cost of a draft on Paris for 30,000 francs, when exchange is at the rate of $1 for 5.25 francs, and gold is worth 106?

9. How large a bill on Paris can be bought for $1,000 in gold, when exchange is at the rate of $1 to 5.20 francs?

REVIEW QUESTIONS.

(**219.**) What is exchange? (**220.**) What is a draft or bill of exchange? A sight draft? A time draft? (**221.**) What is the drawer? The drawee? The payee? A remitter or buyer? An indorser? A holder? (**222.**) What is an acceptance? How made? An acceptor? (**223.**) What is an inland bill? (**224.**) Define par of exchange. Illustrate. (**225.**) What is the course of exchange? Illustrate. (**226.**) What is inland or domestic exchange? How are problems in inland exchange solved? (**228.**) What is foreign exchange? (**229.**) How is foreign exchange computed?

XI. EQUATION OF PAYMENTS.
DEFINITIONS.

230. Equation of Payments is the operation of finding the time at which several debts due at different times may be paid without loss of interest to either party.

The time at which the payment may be made is called the **Equated Time**, and the corresponding date is called the **Equated Date**.

231. The time that a debt has to run is called its **Time of Credit**, and the date from which this time is reckoned is called the **Initial Date**.

A credit of $1 for a unit of time is called a **Unit of Credit**; the product of a debt by its time of credit is its **Amount of Credit**.

The unit of time may be 1 *day*, or 1 *month*, but it must always be the same throughout the same problem.

232. The solution of every problem in *Equation of Payments* depends on the following

PRINCIPLE.

The amount of credit of the sum of several debts is equal to the sum of the amounts of credit of the debts taken separately.

There may be three cases : 1°, the debts may be on one side, and have the same initial date ; 2°, the debts may be on one side, and have different initial dates ; 3°, some of the debts may be on one side, and some on the other, the dates of payment being different.

233. When the Debts are on one side and have the same Initial Date.

On the 1st of January, 1877, A. owes B. $400 payable in 9 months, $800 payable in 6 months, and $600 payable in 4 months; at what date may the whole debt be paid without loss to either party ?

EXPLANATION.—In this case the unit of credit is $1 for 1 month. Multiplying the different sums by their terms of credit and adding the products, we find the sum of the amounts

OPERATION.

$$400 \times 9 = 3600$$
$$800 \times 6 = 4800$$
$$600 \times 4 = 2400$$
$$1800) \qquad 10800 (6$$

Ans. $\begin{cases} \text{Equated time} = 6 \text{ months.} \\ \text{Equated date, July 1, 1874.} \end{cases}$

of credit equal to 10,800 units ; but this is equal to the amount of credit of $1,800 for the required time ; hence, 10,800 divided by 1,800 will give the number of months in the equated time. Adding this to the initial date we have the equated date.

Since all similar cases may be treated in the same manner, we have the following

RULE.

I. Multiply each debt by its time of credit and find the sum of the products ; divide this by the sum of the debts and the quotient will be the equated time.

II. Add the equated time to the initial date and the result will be the equated date.

EXAMPLES.

1. A merchant owes $2,400, of which $400 is payable in 6 *mos.*, $800 in 10 *mos.*, and $1,200 in 16 *mos.*; what is the equated time?   *Ans.* 12⅛ *months.*

2. A. owes B. $2,400, of which $800 is payable in 6 months, $600 in 8 months, and $1,000 in 12 months; what is the equated time?   *Ans.* 9 *months.*

3. A merchant owes $300, payable as follows: $80 in 22 days, $100 in 60 days, and $120 in 75 days; what is the equated time?   *Ans.* 56 *days.*

NOTE.—The rule gives 55⅔⅔ days. In such cases we shall neglect the fraction when it is less than ½, and add 1 to the whole number when it is equal to, or greater than ½. If the unit of time is 1 month, the fractional part will be reduced to days at the rate of 30 days per month, and fractional parts of the latter will be treated as just explained.

4. A. owes B. $1,000, payable as follows: $200 at present, $400 at 6 months, and $400 at 15 months; what is the equated time?   *Ans.* 8 *months* 12 *days.*

5. A. owes a sum of money, of which ⅓ is payable at 30 days, ⅓ at 60 days, and the rest at 90 days; what is the equated time for the payment of the whole?

234. When the Debts are on one side, and have different Initial Dates.

A. sells goods to B. on a credit of 60 days, as follows:

> Jan.   14, 1874, a bill of $2,000;
> Feb.   10, "   "   "   " $1,500;
> March 25, "   "   "   " $3,000;

Required the equated time and date of payment.

EXPLANATION.—The first payment is due March 15, the second April 11, and the third May 24. Assuming the date of the earliest payment as the initial date, the other payments will have 27 and 70 days to

OPERATION.

March 15, 2000 × 0 = 00000
April 11, 1500 × 27 = 40500
May 24, 3000 × 70 = 210000

6500) 250500(38.538

Ans. $\begin{cases} \text{Equated time, 39 } days. \\ \text{Equated date, April 23, 1874.} \end{cases}$

run. Proceeding as before, we find the equated time greater than 38.5 days; calling it 39 days, according to the rule, and adding it to the assumed initial date, we find the equated date to be March 15 + 39 *da.* = April 23.

Since all similar cases may be treated in the same manner, we have the following

RULE.

Find the date of each payment, and assume the earliest as an initial date; then find the time of credit of each item, and proceed as in the last case.

EXAMPLES.

1. A. bought goods as follows : May 5, $500 on 4 months' credit; May 25, $750 on 4 months' credit; June 27, $800 on 6 months' credit; what is the equated date of payment of the whole debt? *Ans.* Oct. 26.

NOTE.—When credit is given by months, calendar months are understood, without reference to the number of days they contain; in applying the preceding rule, the times of credit, if not exact months, are reduced to days, counting the actual number of days in each month. Thus, in the example just given, the items mature Sept. 5, Sept. 25, and Dec. 27, and the times of credit, counting from Sept. 5 as the initial date, are 0, 20, and 113 days.

2. There are three notes payable as follows: the *first* for $500, Feb. 12, 1877 ; the *second* for $400, March 12, 1877; and the *third* for $300, April 1, 1877; what is the equated date for the payment of all ? *Ans.* March 5, 1877.

235. When each party owes the other, the times of payment being different.

The difference of the sums of the *debits* and *credits* of an account is called the **Balance of Account.**

The balance must be added to the smaller sum.

By preceding rules, all the items on either side may be reduced to a single item, payable at a specified time. Hence, every account may be reduced to two items, one on each side.

The following suppositions illustrate every case that can arise:

1°. Suppose that A. owes B. $500 payable July 16, 1877, and that B. owes A. $200 payable July 1, 1877.

2°. Suppose that A. owes B. $200 payable July 16, 1877, and that B. owes A. $500 payable July 1, 1877.

In both cases it is required to find the date at which the *balance of account* may be paid without loss to either party.

In each example let the date of the earlier payment, July 1, 1877, be taken as the initial date.

1°. EXPLANATION.—On the 1st of July, A. was entitled to 500 × 15, or 7,500, units of credit; but at that time the balance due B. was but $300; hence, the time before the balance became due was $\frac{7500}{300}$, or 25 days; adding this to the initial date, we find that the balance was due July 26, 1877.

OPERATION.

$$500 \times 15 = 7500$$
$$\frac{7500}{300} = 25.$$

Ans. July 26, 1877.

2°. EXPLANATION.—On the 1st of July, A. was entitled to 200 × 15, or 3,000, units of credit; but B. was also entitled to the same amount; hence, the balance due him, $300, was entitled to credit for $\frac{3000}{300}$, or 10 days *before* the 1st of July, that is, from the 21st of June, 1877.

OPERATION.

$$200 \times 15 = 3000$$
$$\frac{3000}{300} = 10.$$

Ans. June 21, 1877.

Since all similar cases can be treated in the same manner, we have the following

R U L E.

I. Reduce each side to a single term, and take the date of the earlier payment as an initial date; then multiply the side of the account that falls due last by the time between the dates of payment and divide the product by the balance of the account; the quotient will be the equated time.

II. If the greater side of the account falls due last, add the equated time to the initial date; if the smaller side falls due last, subtract the equated time from the initial date; the result will be the equated date at which the balance is payable.

E X A M P L E S.

1. A. owes B. $8,750 payable July 21, 1877, and B. owes A. $6,500 payable June 9, 1877; when, and to whom, is the balance payable? *Ans.* To B. Nov. 19, 1877.

2. A. owes B. $6,500 payable July 21, 1877, and B. owes A. $8,750, payable June 9, 1877; to whom is the balance payable, and what is its equated date?

Ans. To A., and the equated date is Feb. 8, 1877.

EQUATION OF ACCOUNTS.

236. **Equation of Accounts** is the operation of finding the time at which the balance of an account should be made payable in order that there may be no loss of interest to either party. This time is called the **Equated Time.**

237. Let it be required to find the equated time of payment of the balance of the following account:

Dr. HENRY AHL *in acct. with* BENJ. BARROL.   *Cr.*

1873.			1873.		
April 1.	Merchandise.	$375 00	April 20.	Cash . . .	$500 00
" 18.	"	250 00	May 20.	" . . .	185 00
May 25.	"	150 00			

EXPLANATION.—The first side of the account is equivalent to a single item of $775 payable April 17, and the second side is equivalent to $685 payable April 28 (Art. **233**).

By the rule of Art. **235**, we find that the balance is payable 84 days preceding April 17, that is, on the 23d of January; hence, the equated date of the balance is January 23, 1873.

All similar cases may be treated in like manner; hence the following

R U L E .

Equate each side separately by the rule of Art. **234**; *then find the equated time of the balance by the rule of Art.* **235**.

E X A M P L E .

1. Find the equated date of payment of the balance in the following account:

Dr.   A. *in acct. with* B.   *Cr.*

1874.			1874.		
May 1.	To corn . .	$400 00	June 1.	By cash . .	$300 00
June 15.	" wheat .	1800 00	July 1.	" " . .	1500 00
Oct. 15.	" corn . .	1200 00	Nov. 1.	" " . .	1000 00
		$3400 00			$2800 00

Ans. April 25, 1874.

CASH, AND INTEREST BALANCE.

238. The **Cash Balance** of an account is the difference between the two sides of an account with interest on each item to the date of settlement.

The **Interest Balance** is the difference between the total interest on the items of the two sides.

From these definitions we have the following

RULE.

Add to each side of the account the total interest on all the items of that side to the date of settlement; the difference between the footings of the sides so increased is the cash balance.

EXAMPLES.

1. Find the cash balance of the following account on the 15th of August, 1874, interest being at 6%:

Dr. JNO. IRVING, *in acct. with* HENRY HOUR. *Cr.*

1874.			1874.		
April 1.	To mdse. .	$25 00	May 10.	By cash .	$30 00
" 20.	" " . .	18 00	Aug. 12.	" " . .	35 00
June 15.	" " . .	40 00	" 15.	Cash bal.	18 822
Aug. 15.	Bal. of Int. .	822			
		$83 822			$83 822

EXPLANATION.—The debit side of the account is entitled to interest on $25 for 136 days, on $18 for 117 days, and on $40 for 61 days, all of which is equivalent to the interest on $1 for 7,946 days ; in like manner, the credit side is entitled to interest on $1 for 3.015 days. The debit side is therefore entitled to a balance of interest equal to that on $1 for 4,931 days, or to $.822. We add this to the debit side and then find that the sum of the items on that side exceeds the sum of the items on the credit side by $18.822. This is the cash balance and is to be added to the credit side.

The interest on $1 for 4,931 days is found by the rule for days; that is, we divide 4,931 by 6,000, and the quotient is the number of dollars in the required interest.

2. Find the cash balance of the following account on March 20, 1873, interest being computed at 7%:

Dr.   C. *in acct. with* D.   *Cr.*

1873.					1873.			
Feb. 5.	To mdse. . .	$680	00		Jan. 25.	To cash . .	$250	00
" 24.	" " . .	300	00		Feb. 15.	" " . .	600	00
Mar. 1.	" " . .	150	00					
" 16.	" " . .	600	00					

Ans. $881.631.

REVIEW QUESTIONS.

(**230.**) What is equation of payments? The equated time? The equated date? (**231.**) What is the time of credit? The initial date? What is a unit of credit? The amount of credit? (**232.**) On what principle does equation of payments depend? How many cases may arise, and what are they? (**233.**) What is the rule when the debts are on one side and have the same initial date? (**234.**) Rule when the debts are on both sides and have different initial dates? (**235.**) Rule when each party owes the other, the times of payment being different? (**236.**) What is equation of accounts? Rule? (**238.**) What is the cash balance? The interest balance? Rule for finding cash balance?

XII. CUSTOM-HOUSE BUSINESS.

DEFINITIONS.

239. Duties are taxes laid by the United States government on certain kinds of imported goods.

The ports at which foreign goods may be landed are called **Ports of Entry**. At each port of entry is a *custom-house*, at which the taxes are *collected*.

240. Duties are of two kinds, *specific* and *ad valorem*. A **Specific Duty** is a definite tax laid on a certain article without reference to its cost. An **Ad Valorem Duty** is a percentage on the invoiced price of an article at the place from which it is imported.

241. An **Invoice** or **Manifest** is an inventory of goods setting forth their cost price at the place from which they are imported.

242. A **Tonnage Duty** is a tax laid on a vessel for the privilege of entering a port. This tax depends on the size, or tonnage, of the vessel, the ton being equal to a capacity of 40 cubic feet.

Weights of articles are generally expressed in tons of 2,240 *lbs.* The weight of an article, before any allowance is made, is called its **Gross Weight** ; its weight, after deducting all allowances, is called its **Net Weight**.

243. In computing specific duties certain allowances are made for losses incidental to commerce.

Draft is an allowance for general waste; it is deducted before any other allowance is made. **Tare** is an allowance for the weight of the box, or case, containing goods. **Leakage** is an allowance for loss on liquors imported in casks. **Breakage** is an allowance for loss on liquors imported in bottles.

EXAMPLES.

1. What is the net weight of 176 *hhds.* of sugar, each weighing 10 *cwt.* 1 *qr.* 14 *lbs.*, draft being 7 *lbs.* per *cwt.* and tare 56 *lbs.* per *hhd.* ?   *Ans.* 1,623 *cwt.* 3 *qrs.* 14 *lbs.*

EXPLANATION.—In computing the weight, it is to be remembered that 28 *lbs.* make 1 *qr.*, 4 *qrs.* make 1 *cwt.*, and 20 *cwt.* make 1 *ton.* In this case the gross weight is 1,826 *cwt.*; the draft is 114 *cwt.* 14 *lbs.*; and the tare is 88 *cwt.*

9

2. What is the net weight of 100 tierces of rice, each weighing 250 *lbs.*, tare being 5%?   *Ans.* 23,750 *lbs.*

3. A merchant imported goods invoiced as follows:

300 *yds.* cloth @ 13*s.* per *yd.*;
873 *yds.* carpeting @ 4*s.* per *yd.*;
150 *doz.* cambric hdkfs. @ 2*s.* 8*d.* per dozen.

The duty on the cloth was 30%; on the carpeting 25%; and on the hdkfs. 33%; what was the amount of duty in dollars, allowing $4.8665 to the pound sterling?

Ans. $529.232.

4. A wine merchant imported wines as follows:

75 baskets champagne @ $12 per basket;
18 casks Madeira @ $45 per cask; .
12 casks sherry @ $35 per cask.

Allowing 4% for leakage on the wine in casks, what is the amount of duty @ 45%?   *Ans.* $936.36.

5. What is the duty on 54 *T.* 13 *cwt.* 3 *qrs.* 20 *lbs.* of iron, invoiced at $45 per ton, the duty being $33\frac{1}{3}$%?

Ans. $820.446.

REVIEW QUESTIONS.

(**239.**) What are duties? A port of entry? (**240.**) What is a specific duty? An ad valorem duty? (**241.**) What is an invoice or manifest? (**242.**) What is tonnage duty? Gross weight? Net weight? (**243.**) What is draft? Tare? Leakage? Breakage?

RATIO AND PROPORTION.

I. RATIO.

DEFINITIONS.

244. The **Ratio of one number to another** is the quotient of the *second* number divided by the *first*. Thus, the ratio of 3 to 15 is 15 ÷ 3, or 5.

The first number is called the **Antecedent**, the second number is called the **Consequent**, and both are called **Terms** of the ratio. Thus, in the ratio just given, 3 is the *antecedent*, 15 is the *consequent*, and both 3 and 15 are terms of the ratio.

MENTAL EXERCISES.

1. How many quarts in a gallon ? What is the quotient of 1 gallon divided by 1 quart ? In this case, what is the antecedent ? The consequent ? The ratio ?

EXPLANATION.—In measuring anything, the unit of measure, which is supposed to be known, is the *antecedent ;* the thing to be measured, or expressed in terms of this unit, is the *consequent ;* and the number of times that the antecedent is contained in the consequent is the *ratio.*

2. If we take 3 *inches* as a unit, what is the measure of 1 foot ? How many times 3 inches in 1 foot ? What

then is the ratio of 3 inches to 1 foot ? In this case, what is the antecedent ? What is the consequent ? What is the ratio ?

3. How many times is 3*s.* 4*d.* contained in £1 ? What is the ration of 3*s.* 4*d.* to £1 ? What is the antecedent, or unit ? What is the consequent, or measure ? What is the ratio ?

· METHODS OF EXPRESSING A RATIO.

245. A ratio may be written in the form of a fraction, or it may be expressed by writing the consequent after the antecedent with a *colon* between them. Thus, the ratio of 2 to 4 may be written $\frac{2}{4}$, or it may be written 2 : 4 ; in either case it may be read : *the ratio of 2 to 4*, or, *2 is to 4.*

NOTE.—Because a ratio is a fraction we may multiply or divide both of its terms without altering its value.

246. From the definition of a ratio, we have the following relations :

1°. The ratio = The consequent ÷ The antecedent.

2°. The consequent = The antecedent × The ratio.

3°. The antecedent = The consequent ÷ The ratio.

EXAMPLES

What is the ratio

1. Of 3 to 5 ?
2. Of 7 to 35 ?
3. Of 3*s.* 6*d.* to 17*s.* 6*d.?*
4. Of 14$\frac{3}{8}$ *lbs.* to 43$\frac{1}{8}$ *lbs.?*
5. Of 27$\frac{1}{4}$ *bu.* to 6$\frac{13}{16}$ *bu.?*
6. Of 3$\frac{5}{12}$ *da.* to 34$\frac{1}{8}$ *da.?*
7. Of $4 to $120 ?
8. Of 64 *yds.* to 4 *yds.?*
9. Of 72 *cts.* to 9 *cts.?*
10. Of 13*s.* 9*d.* to 2*s.* 9*d.?*

What is the consequent when

11. The antecedent is 7 and the ratio 4 ?
12. The antecedent is $\frac{1}{2}$ and the ratio $\frac{1}{4}$?
13. The antecedent is 5 *lbs.* 4 *oz.*, and the ratio $\frac{1}{2}$?

What is the antecedent when

14. The consequent is 18 *lbs.* 6 *oz.* and the ratio 6 ?
15. The consequent is $12.75 and the ratio 4.25 ?
16. The consequent is 150 *yds.* and the ratio $7\frac{1}{2}$?

REVIEW QUESTIONS.

(**244.**) What is the ratio of one number to another? What is the antecedent? The consequent? Terms? In measuring a magnitude, what is the antecedent? The consequent? The ratio? (**245.**) Explain the different ways of writing and reading a ratio. (**246.**) Give the value of the ratio, the antecedent, and the consequent each in terms of the other two.

II. PROPORTION.

DEFINITIONS.

247. A **Proportion** is an expression of equality between two ratios. Thus, the expression

$$2 : 5 = 8 : 20,$$

is a *proportion.* It indicates that the ratio of 2 to 5 is equal to the ratio of 8 to 20.

In writing a proportion the sign of equality is usually replaced by a double colon. Thus, the preceding proportion may be written

$$2 : 5 :: 8 : 20.$$

It is then read 2 *is to* 5 *as* 8 *is to* 20.

The *first* and *fourth* terms of a proportion are called **Extremes**; the *second* and *third* terms are called **Means**. Thus, in the preceding proportion, 2 and 20 are the *extremes* and 5 and 8 are the *means*.

SOLUTION OF A PROPORTION.

248. The **Solution of a Proportion** is the operation of finding one of its terms, when the other three are known. The rule by which the solution is performed is called **The Rule of Three.**

PRINCIPLES USED IN SOLVING A PROPORTION.

249. If we have the proportion $2 : 5 :: 8 : 20$, we may write it under the form $\frac{5}{2} = \frac{20}{8}$. If we multiply both terms of the first ratio by 8 and both terms of the second ratio by 2, (Art. **245**), we shall have $\frac{5 \times 8}{2 \times 8} = \frac{2 \times 20}{2 \times 8}$.

Now, these fractions are equal, and they have equal denominators; hence, their numerators are equal, that is, $5 \times 8 = 2 \times 20$; in this case the product of the means is equal to the product of the extremes. But we can reason in like manner on any proportion; hence, we have the following principle:

1°. *The product of the means of any proportion is equal to the product of its extremes.*

From this principle we have the two following

2°. *Either extreme is equal to the product of the means divided by the other extreme.*

3°. *Either mean is equal to the product of the extremes divided by the other mean.*

250. In applying the preceding principles to the solution of proportions, it is found convenient to represent the required term by some letter, as x, and the ratio into which this term enters is written after the other.

EXAMPLES.

1. Let it be required to solve the proportion,

$$15 : 45 :: 9 : x.$$

EXPLANATION.—In this case the required term is one of the extremes; hence, from principle 2°, we have,

$$x = 45 \times 9 \div 15 = 27. \; Ans.$$

NOTE.—After indicating the solution, we cancel all factors common to both numerator and denominator.

2. Let it be required to solve the proportion,

$$\$7 : \$13 :: 56\,lbs. : x. \quad Ans. \; x = 104\,lbs.$$

NOTE.—If the first two terms are denominate, we disregard their common unit. Thus, in the last example, the ratio of $7 to $13 is the same as the ratio of 7 to 3.

Find the value of x in each of the following examples:

3. $15 : \$3 :: x : 4\,yds.$

4. $2 : 3 :: 18 : x.$

5. $8 : 32 :: 24 : x.$

6. $32 : 18 :: 16 : x.$

7. $8 : 4 :: \frac{1}{3} : x.$

8. $1.2 : 6 :: x : 1.3.$

9. $5\,ft. : 7\,ft. :: \$3 : x.$

10. $9\,da. : 15\,da. :: \pounds 2.1 : x.$

11. $30\,ft. : 12\tfrac{1}{2}\,ft. :: \$650 : x.$

12. $28 : 1\tfrac{3}{4} :: \$140 : x.$

13. $25 : 14\tfrac{3}{4} :: \pounds 7 \; 10s. : x.$

14. $84.50 : 21.12\tfrac{1}{2} :: 13\,C. : x.$

15. $75 : 4.75 :: x : \$10\tfrac{1}{2}.$

16. $\pounds 1\tfrac{1}{3} : \pounds 7\tfrac{7}{8} :: 7 : x.$

17. $12\,lbs. : 30\,lbs. :: \$2 : x.$

18. $17\,bu. : 43\,bu. :: 25\tfrac{1}{2} : x.$

19. $100\,ft. : 1\,ft. :: \$150 : x.$

20. $36 : 21 :: \$90 : x.$

21. $44 : 40 :: \$23 : x.$

22. $\pounds 7 : \pounds 11\tfrac{1}{2} :: 4\,yds. : x.$

RULE OF THREE.

251. The **Rule of Three** is a rule for finding from *three* numbers a *fourth*, to which the *third* shall have the same ratio that the *first* has to the *second*.

This rule depends on the principles of article **249.**

252. Let it be required to solve the following problem: If 40 *yds.* of cloth cost \$170, what will 64 *yds.* cost?

EXPLANATION.—We first *state the problem;* that is, we express the conditions of the problem in the form of a proportion.

Having written *x* for the *fourth* term, we write the number having the same unit, that is, \$170, for the *third* term. We then consider whether the *fourth* term is greater, or less than the third; here it is plain that 64 yards costs more than 40 yards, that is, the fourth term is greater than the third; we therefore write the smaller of the remaining numbers for the *first* term, and the *greater one* for the second term. Having completed the statement, we solve the resulting proportion as already explained.

Since all similar cases may be treated in like manner, we have the

SOLUTION.

$$40 \, yds. : 64 \, yds. :: \$170 : x.$$

$$\begin{array}{r} 170 \\ 40)\overline{10880} \\ \hline \$272 \quad Ans. \end{array}$$

RULE OF THREE.

I. Denote the required number by x, and write it for the fourth term of a proportion; then write the number that has the same unit for the third term.

II. Consider, from the nature of the question, whether the fourth term will be greater, or less than the third, and write the remaining numbers, in the same relative order, for the first and second terms.

III. Solve the resulting proportion, and the value of x will be the answer.

EXAMPLES.

1. If 25 yards of silk cost \$81.25, what will 37 yards cost?

EXPLANATION.—The fourth term x stands for the cost of 37 *yds.*; the third term is therefore the cost of 25 *yds.*, that is, $81.25. Now, the cost of 37 *yds.* is greater than that of 25 *yds.*; hence, the first term must be less then the second. We have therefore the following

STATEMENT.—25 *yds.* : 37 *yds.* : : $81.25 : x.
Solving the proportion, we have $x = $120.25. *Ans.*

2. If a man can walk 83 miles in 3 days, how far can he walk in 11 days?

STATEMENT.—3 *days* : 11 *days* : : 84 *miles* : x;
∵ $x = 308$ *miles*. *Ans.*

NOTE.—The sign ∴ stands for *hence*.

3. If it costs £2 9*s.* 6*d.* to travel 198 miles, how far can I travel for £8 0*s.* 10½*d.*?

STATEMENT.—£2 9*s.* 6*d.* : £8 0*s.* 10½*d.* : : 198 *mi.* : x.

4. If 12¼ *lbs.* of gold cost $2,878.75, what will 3 *oz.* cost?

5. If 31 *cwt.* 1 *qr.* 14 *lbs.* of sugar cost $318.45, what will 1,240 *lbs.* cost?

NOTE.—If all the numbers have the same unit, the nature of the question will show which is to be the third term.

6. If a piece of property worth $3,250 is taxed $35.75, what should be the tax on a house worth $17,350?

EXPLANATION.—Here the answer is to be the tax on $17,350; hence, the third term must be the tax on $3,250.

STATEMENT.—$3,250 : $17,350 : : $35.75 : x.

7. Solve the proportion 3 : 4 : : 21 : x.

8. If 3 pairs of socks cost $1.41, what will 7 pairs cost?

9. If 4½ tons of hay will keep 2 cows for the winter, how many cows can be kept on 24¾ tons?

10. If 18¾ bags of coffee contain 758 *lbs.* 8 *oz.*, how many bags are there in 12,136 *lbs.*?

11. How long will it take to travel 1,290 miles, at the rate of 306¾ miles in 20¼ days?

12. If 24 *yds.* 3 *qrs.* of carpeting, 1 yard wide, will cover a room, how many *yds.* of carpeting, 1¼ *yds.* wide will it take to cover the same room?

EXPLANATION.—It will take fewer yards of the latter width than of the former; hence, the fourth term is less than the third.

STATEMENT.—1¼ *yds.* : 1 *yd.* : : 24 *yds.* 3 *qrs.* : *x.*

13. If 12 men can build a wall in 20 days, how many men would it take to build it in 5 days?

EXPLANATION.—It requires more men to do it in 5 than in 20 days; hence, the fourth term is greater than the third.

STATEMENT.—5 *da.* : 20 *da.* : : 12 *men* : *x.*

14. If a piece of cloth 20 yards long and ¾ of a yard wide is required to make a dress, what must be the width of a piece 12 yards long to make the same dress?

STATEMENT.—12 : 20 : : ¾ *yd.* : *x.*

15. In what time can 25 men do a piece of work that 12 men can do in 3 days?

STATEMENT.—25 *men* : 12 *men* : : 3 *days* : *x.*

16. A. exchanged 60 yards of silk, worth $2.40 per yard, for 48 yards of velvet; what did the velvet cost per yard?

17. If 30 bushels of oats, at 50 cents a bushel, will pay a debt, how much barley must be given to pay the same debt, barley being worth 75 cents a bushel?

18. If 42 tons of coal cost $197.40, what will 1¾ tons cost?

19. If a man can do a piece of work in 20 days, working 10 hours per day, how long will it take him to do the same if he works 12 hours per day?

20. If 17⅜ cords of wood cost $88⅝, what will 254¾ *cu. ft.* cost ?

21. If 7₁⁷₁ barrels of apples cost $31⅓, what will 32⅝ barrels cost ?

22. If 2 *bu.* 1 *pk.* of wheat cost $1.93¾, how many bushels can be bought for $96⅞ ?

23. If 10 bushels of coal cost 25.50 *fr.*, what will 13 bushels cost ?

24. If 14 meters of cloth can be bought for 350 *fr.*, how many meters can be bought for 875 *fr.?*

25. If it cost $40 to board 3 men 5 weeks, what will it cost to board 12 men 10 weeks ?

EXPLANATION.—The board of 3 men for 5 weeks is the same as the board of 1 man for 15 weeks, and the board of 12 men for 10 weeks is the same as the board of 1 man for 120 weeks ; hence, the following

STATEMENT.—15 *wks. board* : 120 *wks. board* : : $40 : *x.*

26. If 18 men men consume 34 *bbls.* of potatoes in 135 days, how long will it take 45 men to consume 102 *bbls.?*

EXPLANATION.—In the first case each man consumes $\frac{34}{18}$ *bbls.* and in the second case $\frac{102}{45}$ *bbls. :* in the first case the time is 135 *da.* and in the second case it is *x da.* Now, it will take longer to consume $\frac{102}{45}$ *bbl.* than to consume $\frac{34}{18}$ *bbl.*, that is, the fourth term is greater than the third ; hence, the

STATEMENT.—$\frac{34}{18}$ *bbl.* : $\frac{102}{45}$ *bbl.* : : 135 *da.* : *x.*

27. If 12 boys pay $2,000 for 1 year's tuition, what must 14 boys pay for 18½ months' tuition ?

STATEMENT.—12 × 12 *mo.* : 14 × 18½ *mo.* : : $2,000 : *x.*

28. If it costs $7.20 to transport 18½ *cwt.*, 5½ *mi.*, what will it cost to transport 112¾ *T.*, 62½ miles?

29. If 20 men working 11 hours a day for 30 days can

earn $3,300, how much can 36 men earn in 40 days working 10 hours per day?

STATEMENT.—20 . 30 . 11 : 36 . 40 . 10 : : $3,300 : x.

30. If 7 men reap 6 acres in 12 hours, how many men must be employed to reap 15 acres in 14 hours?

STATEMENT.—$\frac{6}{12}$ A. : $\frac{15}{14}$ A. : : 7 men : x.

$$\therefore x = 15 \text{ men.} \quad Ans.$$

31. If 14 horses eat 56 bushels of oats in 16 days, how many horses will it take to eat 120 bushels in 24 days?

STATEMENT.—$\frac{56}{16}$ bu. : $\frac{120}{24}$ bu. : : 14 horses : x.

$$\therefore x = 20 \text{ horses.} \quad Ans.$$

32. If 12 horses will plow 11 A. in 5 days, how many horses will be required to plow 33 A. in 18 days?

33. If a man can walk 250 miles in 9 days of 12 hours each, how many days of 10 hours each would it take him to walk 400 miles.   *Ans.* 17$\frac{7}{25}$ *days.*

DISTRIBUTIVE PROPORTION AND PARTNERSHIP.

253. The *rule of three* enables us to divide a number into parts proportional to two or more given numbers.

EXAMPLES.

1. Let $140 be divided into three parts, proportional to 3, 5, and 6.

EXPLANATION.—The sum of the numbers to which the parts are proportional is 14. Now, $140 bears the same relation to the first part that 14 bears to 3.

Hence, 14 : 3 : : $140 : *the first part.*
Also, 14 : 5 : : $140 : *the second part.*
And, 14 : 6 : : $140 : *the third part.*

Solving these proportions, we have,

1st part, $30 ; 2d part, $50 ; 3d part, $60.

2. Divide 20 *lbs.* 4 *oz.* into 3 parts proportional to 3, 5, and 10.   *Ans.* 3 *lbs.* 6 *oz.; 5 lbs.* 10 *oz.;* and 11 *lbs.* 4 *oz.*

3. Divide $540 among a man, his wife, and three children, so that the wife shall have twice as much as each child, and the man twice as much as his wife.

Ans. Man, $240; wife, $120; child, $60.

4. A. and B. start from places 150 miles apart and travel towards each other; A. travels 7 *mi.* per hour,. and B. travels 8 *mi.* per hour; how far does each travel before they meet?

Ans. A 70 *mi.;* B. 80 *mi.*

5. A., B., and C. enter into partnership; A. puts in $720, B. puts in $340, and C. puts in $960; if they gain $505, how much should each receive?

6. A. and B. buy goods to the amount of $400, of which A. pays $150 and B. $250; if they lose $100, how much of the loss must each bear?

7. A., B., and C. engage in a speculation towards which A. contributes $480, B. $720, and C. $1,200; if they all gain $650, how much does each gain?

8. A bankrupt owes A. $500, B, $750, C., $900, and D., $1,250, but his estate is worth only $1,020; what share ought each to receive?

ANALYSIS.

254. **Analysis** is the method of solving problems by the direct application of general principles, without the use of particular rules.

Many of the problems usually solved by the rule of three can be solved, more expeditiously, by analysis. The method of proceeding will be shown best by examples.

EXAMPLES.

1. If 34 men can build a house in 40 days, how long will it take 12 men to build the same house?

ANALYSIS.—It will take 1 man 34 times as long as it will 34 men; hence, it will take 1 man 34 × 40,

OPERATION.

$$\frac{34 \times 40 \quad a.}{12} = 113\tfrac{1}{3} \, da. \quad Ans.$$

or 1360 days. But 12 men can build it in $\tfrac{1}{12}$ of the time that 1 man can; hence, 12 men can build it in 1360 $da. \div 12$, or in $113\tfrac{1}{3}$ *days*.

2. If 2 *cwt.* 3 *qrs.* 10 *lbs.* of sugar cost \$34.20, how much can be bought for \$75.60?

ANALYSIS.—If we divide \$34.20 by 285, the quotient, 12 *cts.*, will be the cost of 1 *lb.* If we divide \$75.60 by 12 *cts.*, the quotient will be the required number of pounds. The entire operation is indicated in the last line.

OPERATION.

$$\frac{\$34.20}{285} = 12 \, cts., \text{ cost of 1 } lb.$$

$$\frac{\$75.60}{12 \, cts.} = 630, \, number \, of \, lbs.$$

$$\therefore \frac{75.60 \times 285 \, lbs.}{34.20} = 630 \, lbs. \quad Ans.$$

3. If 6 men can reap 80 acres in 12 days, how many days will it take 25 men to reap 200 acres?

ANALYSIS.—If it takes 6 men 12 days to reap 80 acres, it will take 1 man 6 × 12 days to do the same; hence, 1 man can reap 80 ÷ (6 × 12) acres in 1 day, and, consequently, 25 men can reap 25 times as much in 1 day, that is, they can reap 80 × 25 ÷ (6 × 12) acres; now, if we divide 200 *acres* by the number of acres that 25 men can reap in one day, the quotient, $7\tfrac{1}{2}$, will be the required number of days.

OPERATION.

$$200 \div \frac{80 \times 25}{6 \times 12}$$

$$= \frac{200 \times 6 \times 12}{80 \times 25} = 7\tfrac{1}{2};$$

$$\therefore \; 7\tfrac{1}{2} \text{ days.} \quad Ans.$$

4. A. can do a piece of work in 4 days, and B. can do it in 6 days; how long will it take them to do it, if they work together?

EXPLANATION. — A. can do $\frac{1}{4}$ of the work in 1 day, and B. can do $\frac{1}{6}$ of it in the same time; hence, both together can do $\frac{1}{4} + \frac{1}{6}$, or $\frac{5}{12}$ of it in 1 day; but if they can do $\frac{5}{12}$ of it in 1 day, they can do $\frac{1}{12}$ of it in $\frac{1}{5}$ of 1 day, and, consequently, they can do $\frac{12}{12}$ of it in $\frac{12}{5}$ days, that is, in $2\frac{2}{5}$ days.

OPERATION.

$$\frac{1}{4} + \frac{1}{6} = \frac{5}{12};$$

$$1 \div \frac{5}{12} = 2\frac{2}{5};$$

$$\therefore 2\frac{2}{5} \text{ days. } Ans.$$

5. Three men hire a pasture for $45; the first puts in 3 horses for 5 weeks, the second puts in 4 horses for 3 weeks, and the third puts in 7 horses for 4 weeks; what should each man pay?

EXPLANATION.—Since the pasturage of 3 horses for 5 weeks is equivalent to the pasturage of 1 horse for 15 weeks, the first received the benefit of 15 weeks' pasturage

OPERATION.

1st. $3 \times 5 = 15$; $\frac{15}{55}$ of $45 = $12\frac{3}{11}$;

2d. $4 \times 3 = 12$; $\frac{12}{55}$ of $45 = $9\frac{9}{11}$;

3d. $7 \times 4 = 28$; $\frac{28}{55}$ of $45 = $22\frac{10}{11}$;

Sum. $\overline{55}$

for 1 horse; the second, in like manner, received 12 weeks' pasturage for 1 horse; and the third received 28 weeks' pasturage for 1 horse; they all received 55 weeks' pasturage for 1 horse. Hence, the first should pay $\frac{15}{55}$, the second $\frac{12}{55}$, and the third $\frac{28}{55}$ of the rent.

In like manner other problems may be analyzed and solved. Let all the examples in Article **252** be solved by analysis, and also the following:

6. If 2 *bu.* 1 *pk.* of wheat cost $2.43, what will 14½ *bu.* cost?   *Ans.* $15.39.

7. If 14 men can board 1 week for $45.50, how long can 3 men board for $97.50?   *Ans.* 10 *weeks.*

8. If a steamer sails 728 miles in 2¼ days, how far will she sail in 12¼ days?   *Ans.* 3,900 *miles.*

9. If 20 men perform a piece of work in 12 days, how many men will be required to do a piece of work 3 times as great in ⅕ of the time?   *Ans.* 300.

10. If 450 *lbs.* of coffee cost $99, how much will 1,450 *lbs.* cost?   *Ans.* $319.

11. If 27 tons of iron cost $540, how much will 37½ tons cost?   *Ans.* $750.

12. If 17 bushels of wheat are worth $25.50, how much are 29 bushels worth?   *Ans.* $43.50.

13. If 3 dozen of wine cost $28.50, how much will 5½ dozen cost?   *Ans.* $52.25.

14. If 117 bushels of barley cost $105.30, how much will 413 bushels cost?   *Ans.* $371.70.

15. If 36 gallons of molasses cost $32.40, how much can be bought for $105.30?   *Ans.* 117 *gals.*

16. If I pay $3.00 for riding 40 miles in a stage-coach, how far can I ride for $10.42½?   *Ans.* 139 *miles.*

17. If 32 acres of land can be bought for $1,504, how much can be bought for $5,546?   *Ans.* 118 *acres.*

18. Find the cost of 25½ *lbs.* of tea, when 17 *lbs.* can be bought for $15.30.   *Ans.* $22.95.

19. If 11 Irish miles are equal to 14 English miles, what is the length, in English miles, of a road that measures 57 Irish miles?   *Ans.* 72$\frac{6}{11}$ *English miles.*

20. If a staff 4 feet high casts a shadow 6 feet long, what must be the height of a pole that will cast a shadow 58 feet long at the same time?   *Ans.* 38⅔ *ft.*

21. A man paid \$36 to several laborers; to each man he paid \$4, and to each boy \$2; the number of men was equal to the number of boys; how many were there of each ?   *Ans.* 6.

22. If 14 men consume \$20 worth of flour in 15 days, how many days will it last 21 men ?   *Ans.* 10 *days.*

23. If a barrel of flour will make 180 ten-cent loaves, how many eight-cent loaves will it make ?   *Ans.* 225.

24. A horse and saddle together were worth \$100, and the horse was worth 9 times as much as the saddle; what was the horse alone worth ?   *Ans.* \$90.

25. A farmer puts a flock of sheep in 3 pastures; in the first he puts $\frac{1}{3}$ of his flock, in the second $\frac{1}{4}$ of his flock, and in the third he puts 32 sheep; how many sheep has he ?   *Ans.* 192.

26. What number is that to which if its sixth part and its eighth part be added the sum will be 186 ?

Ans. 144.

27. A woman buys eggs at the rate of 3 for 5 *cts.*, and sells them at the rate of 4 for 7 *cts.*, clearing 9 *cts.* by the bargain; how many does she buy ?   *Ans.* 108.

28. A farmer gave 5 loads of straw for 12 tons of coal, worth \$6$\frac{1}{4}$ per ton; what did he get per load for his straw ?   *Ans.* \$15$\frac{3}{5}$.

29. At what time between 1 and 2 o'clock are the hands of a watch together ?

EXPLANATION.—The hands are together at 12 o'clock, and the hour hand gains 55 minute spaces in an hour; but it must gain 60 minute spaces before they can be together again; hence, the time required is $\frac{60}{55}$ *hrs.*, or 1 *hr.* 5$\frac{5}{11}$ *min.*, that is, they are together at 5$\frac{5}{11}$ minutes past 1. *Ans.*

30. If sugar is worth 7½ *cts.* per lb., how many pounds can be bought for 2½ *T.* of iron, at $60 per ton ?

Ans. 2,000.

NOTE.—Problems relating to the distribution of loss or gain among partners may be solved like example 5.

31. Two men enter into partnership; the first puts in $60, the second $80, and they gain $35; what is the gain of each ?   *Ans.* 1st, $15 ; 2d, $20.

32. A., B., and C. enter into speculation ; A. puts in $4,000, B. puts in $5,000, and C. puts in $6,000; they lose $2,000; what part of the loss must each bear ?

Ans. A., $533.33⅓; B., $666.66⅔; and C., $800.

REVIEW QUESTIONS.

(**247.**) What is a proportion ? How is a proportion written ? How read ? (**248.**) What is the solution of a proportion ? (**249.**) What principles are used in solving proportions ? (**250.**) By what do we represent the unknown term of a proportion ? (**251.**) What is the rule of three ? · (**252.**) Give the rule of three. (**253.**) How are problems in partnership solved ? (**254.**) What is analysis ?

Powers, Roots and Progressions.

I. POWERS.

DEFINITIONS.

255. A **Power** is the product of two or more equal factors. One of these factors is called the **Root** of the power.

The product of two equal factors is called a second power, or square; the product of three equal factors is a third power, or cube; the product of four equal factors is a fourth power; and so on. Thus, 3×3, or 9, is the square of 3; $3 \times 3 \times 3$, or 27, is the cube of 3, and so on

256. The **Exponent** of a power is a number that shows how many times the root is taken as a factor.

It is written to the right and above the root. Thus, in the expression 3^4, 4 is the *exponent;* it indicates that 3 is to be taken 4 times as a factor, that is, $3^4 = 3 \times 3 \times 3 \times 3 = 81$.

INVOLUTION, OR RAISING TO POWERS.

257. Involution is the operation of finding any power of a number.

The operation of involution is also called *raising to powers.*

258. It follows from the definitions already given that we may find any power of a number by the following

R U L E .

Take the number as a factor as many times as there are units in the exponent of the power.

NOTE.—To raise a simple fraction to any power, raise each term separately to that power.

E X A M P L E S .

Raise the following numbers to the powers indicated :

1. $(4)^3$.	6. $(.09)^2$.	11. $(\frac{3}{4})^3$.
2. $(5)^4$.	7. $(.15)^2$.	12. $(\frac{2}{3})^4$.
3. $(14)^3$.	8. $(2.5)^3$.	13. $(2\frac{1}{2})^3$.
4. $(25)^2$.	9. $(.33)^3$.	14. $(3\frac{1}{4})^3$.
5. $(98)^2$.	10. $(3.4)^2$.	15. $(4\frac{1}{2})^4$.

R E V I E W   Q U E S T I O N S .

(**255.**) What is a power? A second power? A third power? (**256.**) What is an exponent? How written? What does it show? (**257.**) What is involution? What other name has it? (**258.**) Rule for raising a number to a power.

------◆------

I I .   R O O T S .

DEFINITIONS.

259. A **Root** of a number is one of its equal factors.

If a number can be resolved, or separated, into two equal factors, it is said to be a *perfect square ;* if it can be resolved into three equal factors, it is said to be a *perfect cube,* and so on.

260. The **Square Root of a Number** is one of its *two* equal factors. Thus, 4 is the square root of 16.

If the number is not a perfect square, its square root is only approximate.

All the perfect squares less than 100, with their square roots, are written in the following

TABLE.

Perfect squares, 1 4 9 16 25 36 49 64 81,
Square roots,   1 2 3 4 5 6 7 8 9.

NOTE.—The sign $\sqrt{}$, called the **Radical Sign**, shows that the square root of the number under it is to be taken. Thus, $\sqrt{36}$ denotes that the square root of 36 is to be taken.

METHOD OF EXTRACTING A SQUARE ROOT.

261. The method of finding the square root of a number depends on the principles of Algebra. (See Manual of Algebra, Art. **107.**) In accordance with these principles, we have the following

RULE.

I. Separate the given number into periods of two figures each, beginning at the right hand; the period on the left will often contain but one figure.

II. Find the greatest perfect square in the first period on the left and place its square root on the right, after the manner of a quotient in division; subtract the square of this root from the first period, and to the remainder bring down the second period for a dividend.

III. Double the root found and place it on the left for a divisor. See how many times this divisor is contained in the dividend, exclusive of the right hand figure, and place the quotient in the root and also at the right of the divisor.

IV. Multiply the divisor, thus augmented, by the last figure of the root already found, subtract the product from the dividend and to the remainder bring down the next period for a new dividend.

V. Double the root already found for a new divisor, and continue as before, until all the periods have been brought down and operated on.

NOTES.—1. If any quotient figure proves too large, let it be diminished until it gives a product less than the partial dividend.

2. If the last remainder is 0, the given a number is a perfect square and the root is exact ; if not, the root is true to within less than 1.

3. The square root of a simple fraction is equal to the square root of its numerator divided by the square root of its denominator.

EXAMPLES.

1. Find the square root of 8836.

EXPLANATION.—The two periods are 88 and 36 ; the greatest perfect square in 88 is 81, (table, **Art. 260**), and its square root is 9 ; this we write as the first figure of the root and place its square 81 under the first period ; subtracting, we have 7 for a remainder, to which we bring down the period 36 for a dividend ; doubling 9 we have 18, which we place on the left for a divisor, and this is contained 4 times in 73 ; we therefore place 4 on the right of 9 and also on the right of 18 ; multiplying 184 by 4 we find 736, which taken from 736 gives 0 for a remainder ; hence, the square root of 8836 is 94.

OPERATION.

$$8836(94$$
$$81$$
$$18{,}4)73{,}6$$
$$73\ 6$$
$$0$$

Perform the following indicated operations :

2. $\sqrt{9604}$.	6. $\sqrt{14641}$.	10. $\sqrt{\frac{121}{144}}$.
3. $\sqrt{13225}$.	7. $\sqrt{37636}$.	11. $\sqrt{\frac{25}{729}}$.
4. $\sqrt{342225}$.	8. $\sqrt{41616}$.	12. $\sqrt{\frac{169}{288}}$.
5. $\sqrt{944784}$.	9. $\sqrt{52441}$.	13. $\sqrt{\frac{25}{361}}$.

NOTE.—If there is a remainder the operation may be continued by annexing periods of decimal ciphers; for each period thus annexed there will be one decimal figure in the root. Thus, $\sqrt{187} = \sqrt{187.0000} = 13.67$. Here the approximate root is true to within less than .01.

Find the square roots of the following numbers to two decimal places:

14. 229.   *Ans.* 15.13.   16. 450.   *Ans.* 21.21.

15. 354.   *Ans.* 18.81.   17. 592.   *Ans.* 24.33.

NOTE.—The square root of a decimal may be found by the preceding rule. In this case we begin to point off periods at the decimal point and proceed toward the right. Any simple fraction may be changed to a decimal and then operated upon by the rule.

Find the square roots of the following numbers to three places of decimals:

18. .0249.   *Ans.* .157.   21. .152881.   *Ans.* .391.

19. .69.   *Ans.* .830.   22. .326041.   *Ans.* .571.

20. .1051.   *Ans.* .324.   23. .010404.   *Ans.* .102.

PRACTICAL PROBLEMS.

1. A general forms an army of 117,649 men in a square; how many men are there in each rank and how many ranks in the square?   *Ans.* 343.

2. In a square pavement there are 48,841 stones, each 1 *ft.* square; what is the length of the pavement and what is its breadth?   *Ans.* 221 *ft.*

3. A square farm contains 640 acres; how long is each side?   *Ans.* 320 *rods.*

4. A square field contains 160 acres; what will it cost to build a wall around if each rod of wall cost $2?

Ans. $1,280.

CUBE ROOT.

262. The **Cube Root of a Number** is one of its *three* equal factors. Thus, 5 is the cube root of 125.

If a number is not a perfect cube its cube root is only approx-imate.

All the perfect cubes less than 1,000, with their cube roots, are written in the following

TABLE.

Perfect cubes,	1	8	27	64	125	216	343	512	729;
Cube roots,		1	2	3	4	5	6	7	8	9.

NOTE.—The sign, $\sqrt[3]{}$, shows that the cube root of the number under it is to be taken. Thus, $\sqrt[3]{125}$ denotes that the cube root of 125 is to be taken. The number 3 written over the sign is called an Index.

METHOD OF EXTRACTING A CUBE ROOT.

263. The method of finding the cube root of a number depends on the principles of algebra, (see Manual of Algebra, Art. **111**). In accordance with these principles we have the following

RULE.

I. Separate the number into periods of three fig-ures each, beginning at the right; the left-hand period will often contain less than three figures.

II. Find the greatest perfect cube in the first period on the left, and set its root on the right after the manner of a quotient in division; subtract the cube of this root from the first period and to the remainder bring down the first figure of the next period for a dividend.

III. Take three times the square of the root thus found for a divisor, find how many times it is contained in the dividend, and place the quotient for a second figure of the root. Cube the number thus found, and, if its cube is greater than the first two periods, diminish it successively by 1 until its cube is less than the first two periods; then subtract the result from the first two periods and to the remainder bring down the first figure of the next period for a new dividend.

IV. Take three times the square of the root found for a new divisor and proceed as before, continuing the operation till the periods have been operated on.

NOTES.—1. If the last remainder is 0 the number is a perfect cube and the root is exact; if not, the root is true to within less than 1.

2. The cube root of a simple fraction is equal to the cube root of its numerator divided by the cube root of its denominator.

3. The cube root of a decimal or the approximate cube root of an imperfect cube, may be found by a process entirely similar to that employed in finding the square root in similar cases.

E X A M P L E S.

1. Find the cube root of 804357.

EXPLANATION.—The number having been separated into periods, we find the greatest cube in 804 to be 729 and its cube root, 9, is the first figure of the root; taking 729 from 804 and bringing down 3, we have 753 ; dividing this by 3 times the square of 9, or 243, we get 3 for the second figure of the root ; cubing 93 we find the result equal to the given number ; hence 93 is the required root.

OPERATION.

$$804\ 357\ (\ 93$$
$$9^3 = 729$$
$$3 \times 9^2 = 243\ )\ 753$$
$$93^3 = 804357$$
$$0$$

Perform the following indicated operations:

2. $\sqrt[3]{531441}$.   *Ans.* 81.   6. $\sqrt[3]{67}$.   *Ans.* 4.06.

3. $\sqrt[3]{970299}$.   *Ans.* 99.   7. $\sqrt[3]{104}$.   *Ans.* 4.7.

4. $\sqrt[3]{35937}$.   *Ans.* 33.   8. $\sqrt[3]{206}$.   *Ans.* 5.9.

5. $\sqrt[3]{224755712}$. *Ans.*608.   9. $\sqrt[3]{585}$.   *Ans.* 8.36.

REVIEW QUESTIONS.

(**259.**) What is a root of a number? What is a perfect square? A perfect cube? (**260.**) What is the square root of a number? What is the radical sign? What does it show? (**261.**) What is the rule for extracting the square root of a number? Of a simple fraction? How do you find an approximate value of a square root? How do you find the square root of a decimal to any number of places? Of a simple fraction reduced to a decimal? (**262.**) What is a cube root? Its sign? What is an index? (**263.**) Give the rule for extracting the cube root of a number.

———◆———

III. PROGRESSIONS.

DEFINITIONS.

264. A **Progression** is a series of numbers that increase, or decrease, according to a common law.

The numbers forming a progression are called **Terms**; the first and last terms are called *extremes* and all the rest are called *means*.

NOTE.—Progressions are of two kinds, *arithmetical* and *geometrical*.

1°. ARITHMETICAL PROGRESSION.

DEFINITIONS.

265. An **Arithmetical Progression** is one in which each term, after the first, is equal to the preceding term *increased*, or *diminished*, by a given number. This number is called the **Common Difference**.

If a progression is formed by the continued addition of a common difference it is *increasing ;* if it is formed by the continued subtraction of a common difference it is *decreasing.*

The first of the following progressions is *increasing* and the second is *decreasing :*

2   4   6   8   10...........∴*Increasing progression.*
10   8   6   4   2...........*Decreasing progression.*

If the *increasing* progression is inverted, that is, if it is taken in a reverse order, it becomes a *decreasing* progression.

TO FIND ANY TERM.

266. From the preceding definitions it is obvious that we may find any term by the following

RULE.

Multiply the common difference by the number of terms that precede the required term; if the progression is increasing, add the product to the first term; if the progression is decreasing, subtract the product from the first term.

EXAMPLES.

1. The first term of an *increasing* arithmetical progression is 3, and the common difference is 3 ; what is the 9th term ?   *Ans.* $3 + 3 \times 8 = 27$.

2. The first term of a *decreasing* arithmetical progression is 36, and the common difference is 6 ; what is the 5th term ?   *Ans.* $36 - 6 \times 4 = 12$.

3. In an increasing progression the first term is 4, and the common difference is 2 ; what is the 20th term ?

Ans. 42.

4. In a decreasing progression the first term is 45, and the common difference 4 ; what is the 8th term ?

Ans. 17.

TO FIND THE SUM OF THE TERMS.

267. A rule for finding the sum of the terms may be deduced by inverting the progression and proceeding as in the following

OPERATION.

| 3, | 6, | 9, | 12, | 15, | 18 . . . Given progression. |

18,   15,   12,   9,   6,   3 . . . Same inverted.

21 + 21 + 21 + 21 + 21 + 21 . . . Sum of both.

EXPLANATION.—The sum of the terms in both progressions is obviously equal to twice the sum of the terms of the given progression; hence, the sum of the terms of the given progression is $\frac{21}{2} \times 6$, or 63.

Since all similar cases may be treated in the same manner, we have the following

RULE.

Multiply half the sum of the extremes by the number of terms.

EXAMPLES.

1. The first term of a progression is 3, the last term is 27, and the number of terms is 9; what is the sum of the terms ?   *Ans.* 135.

NOTE.—If the last term is not given, it may be found by the rule of Article **266.**

2. The first term of a decreasing progression is 36, the common difference is 6, and the number of terms is 5; what is the sum of the terms ?

Ans. 120.

3. In a decreasing progression, the first term is 45, the common difference is 4, and the number of terms is 8; what is the sum of the terms ?   *Ans.* 248.

4. What is the sum of the natural numbers, 1, 2, 3, &c., up to 99, inclusive? *Ans.* 4,950.

5. The first term of a decreasing progression is 15, the last term is 5, and the number of terms is 6; what is the sum of the terms? *Ans.* 60.

6. The first term of an increasing progression is 15, the common difference is 3, and the number of terms is 6; what is the sum of the terms? *Ans.* 135.

7. What is the sum of the terms of the progression 1, 2, 3, 4, etc., up to 12 inclusive? *Ans.* 78.

8. The first term of an increasing progression is 7, the common difference is 4, and the number of terms is 7; what is the sum of the terms? *Ans.* 133.

2°. GEOMETRICAL PROGRESSION.

DEFINITIONS.

268. A **Geometrical Progression** is one in which each term, after the first, is equal to the preceding term *multiplied* by a given number. This number is called the **Ratio** of the progression.

If the ratio is greater than 1, the progression is *increasing;* if less than 1, the progression is *decreasing.* Thus,

$$2, \qquad 4, \qquad 8, \qquad 16,$$

is an increasing progression, and

$$16, \qquad 8, \qquad 4, \qquad 2.$$

is a decreasing progression. The *ratio* in the first case is 2 and in the second case it is $\frac{1}{2}$; in all cases, the *ratio is equal to the quotient obtained by dividing the second term by the first.*

TO FIND ANY TERM.

269. In accordance with the preceding definitions, we may find any term by the following

RULE.

Raise the ratio to a power whose exponent is the number of terms that precede the required term and multiply the first term by the result.

EXAMPLES.

1. In a progression the first term is 3, and the ratio is 3; what is the 6th term? *Ans.* $3 \times 3^5 = 729$.

2. The first term of a progression is 64, and the ratio is $\frac{1}{2}$; what is the 5th term? *Ans.* $64 \times (\frac{1}{2})^4 = 4$.

3. Find the 10th term of the progression 2, 4, 8, &c.

Ans. $2 \times 2^9 = 1,024$.

4. What is the 5th term of the progression 243, 81, 27, &c.? *Ans.* 3.

TO FIND THE SUM OF THE TERMS.

270. Let it be required to find the sum of 4 terms of the series 2, 8, &c.

$2 + 8 + 32 + 128$............Indicated sum of the terms;

$8 + 32 + 128 + 512$........4 times the sum;

$512 - 2$...................3 times the sum;

$$\frac{512 - 2}{3} = \frac{128 \times 4 - 2}{3} = 170\ldots\text{Required sum.}$$

EXPLANATION.—Having indicated the sum of the terms, we multiply each by 4 and set the products one place toward the right; the sum of these results is 4 times the sum of the given series; subtracting the latter from the former, we find $512 - 2$, which is 3 times the required sum; dividing by 3, we have 170, which is the required sum.

Since all similar cases may be treated in like manner, we have the following

R U L E .

Multiply the last term by the ratio; take the difference between the product and the first term; multiply this by the difference between 1 and the ratio.

E X A M P L E S .

1. The first term of a progression is 3, the last term is 729, and the ratio is 3 ; what is the sum of the terms ? $Ans. \dfrac{729 \times 3 - 3}{2} = 1,092.$

NOTE.—If the first term and ratio are given, the last term may be found by the preceding rule.

2. The first term of a progression is 2, the ratio is 4, and the number of terms is 5 ; what is the sum of the terms ? *Ans.* 682.

3. The first term of a geometrical progression is 3 and the ratio is 2 ; what is the sum of 6 terms ? *Ans.* 189.

4. The first term of a geometrical progression is 64 and the ratio is $\frac{1}{2}$; what is the sum of 6 terms ? *Ans.* 126.

5. What is the sum of 7 terms of the progression 2, 6, 18, etc.? *Ans.* 2,186.

R E V I E W   Q U E S T I O N S .

(**264.**) What is a progression ? What are terms ? Extremes ? Means ? (**265.**) What is an arithmetical progression ? An increasing progression ? A decreasing progression ? (**266.**) What is the rule for finding any term ? (**267.**) For finding the sum of the terms ? (**268.**) What is a geometrical progression ? What is the ratio ? When is the progression increasing and when decreasing ? (**269.**) How do you find any term ? (**270.**) The sum of any number of terms ?

MENSURATION.

271. Mensuration is the operation of finding how many times any given magnitude contains its unit of measure.

The unit of measure of a magnitude is always a magnitude of the same kind. The unit of a line, or the *linear unit*, is a straight line of given length; as *one foot:* the unit of a surface, or the *superficial unit*, is a square whose sides are equal to the linear unit; as *one square foot:* the unit of a volume, or the *cubic unit*, is a cube whose edges are equal to the linear unit; as *one cubic foot.*

NOTE.—The rules for mensuration depend on the definitions and principles of Geometry, some of which have already been given. In what follows, the references are to the **Manual of Geometry**. In these references B. stands for Book and P. for Proposition.

272. A Polygon is a plane figure bounded on all sides by straight lines.

Each of the bounding lines is called a **Side** of the polygon, and the point at which any two sides meet is called a **Vertex** of the polygon.

273. The **Area of a Polygon** is the number of superficial units that it contains.

274. A Triangle is a polygon of three sides; as ABC; the side AB on which it

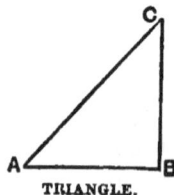

TRIANGLE.

is supposed to stand is called its **Base,** and the shortest distance, CB, from the opposite vertex to the base is called its **Altitude.**

A **Right-angled Triangle** is a triangle that has one right angle. Thus, ABC is a right-angled triangle. The side AC, opposite the right angle, is called the **Hypothenuse.**

275. A Parallelogram is a polygon of four sides, parallel two and two. A **Rectangle** is a right-angled parallelogram.

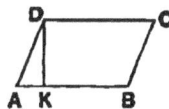

PARALLELOGRAM.   RECTANGLE.

The figure ABCD is a parallelogram whose *base* is AB and whose *altitude,* or *breadth,* is KD; the figure EFGH is a rectangle whose *base* is EF and whose *altitude,* or *breadth,* is FG.

276. A Trapezoid is a polygon of four sides, only two of which are parallel.

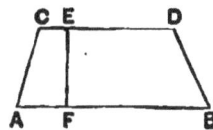

TRAPEZOID.

The figure ABCD is a trapezoid; the longer one of its parallel sides, is its *lower base,* the shorter one its *upper base,* and the perpendicular distance between them is its *altitude.*

277. A Prism is a solid bounded by two parallel polygons called **Bases,** and by parallelograms called **Lateral Faces.**

Prisms are named from their bases. The figures in the margin show a *quadrangular* prism and a *hexagonal* prism.

QUADRANGULAR PRISM.   HEXAGONAL PRISM.

The **Altitude of a Prism** is the shortest distance between its bases.

.10

278. A **Pyramid** is a solid bounded by a polygon called the **Base,** and by three or more triangles called **Lateral Faces.** The lateral faces meet at a point which is called the **Vertex** of the pyramid.

PYRAMID.   FRUSTUM.

If a pyramid is cut by a plane parallel to the base, the part included between this plane and the base is called a **Frustum of a Pyramid.** The **Altitude of a Pyramid** is the shortest distance from the vertex to the base. The **Altitude of a Frustum** is the shortest distance between its bases.

The figure shows a pyramid and a frustum of a pyramid.

279. A **Cylinder** is a solid bounded by two equal and parallel circles called **Bases,** and by a curved surface called the **Convex Surface.**

The **Altitude of a Cylinder** is the shortest distance between its bases.

The figure shows a cylinder.

CYLINDER.

280. A **Cone** is a solid bounded by a circle called the **Base,** and by a curved surface called the **Convex Surface.** The convex surface tapers uniformly from the

CONE.   FRUSTUM.

base to a point which is called the **Vertex of the Cone.**

If a cone is cut by a plane parallel to the base, the part included between this plane and the base is called a **Frustum of a Cone.**

The **Altitude of a Cone** is the shortest distance from the vertex to the base. The **Altitude of a Frustum** is the shortest distance between its bases.

The figure shows a cone and the frustum of a cone.

281. A **Sphere** is a solid every point of whose surface is equally distant from a point within called **The Centre.**

A straight line from the centre of a sphere to any point of the surface is

SPHERE.

called a **Radius.** A straight line through the centre and terminating at both ends in the surface is called a Diameter. A plane through the centre of a sphere cuts from the sphere a **Great Circle.** Any plane that intersects the sphere but does not pass through the centre, cuts from the sphere a **Small Circle.**

PROPERTY OF THE RIGHT-ANGLED TRIANGLE.

282. It is shown in geometry (B. 4., P. 8), that the square of the hypothenuse of a right-angled triangle is equal to the sum of the squares of the other two sides. Calling the sides about the right-angle, the **Base** and the **Altitude,** we have the following relations:

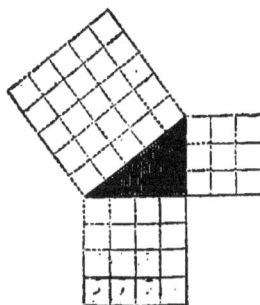

1°. $Hypothenuse = \sqrt{(Base)^2 + (Altitude)^2}.$

2°. $Base = \sqrt{(Hypothenuse)^2 - (Altitude)^2}.$

3°. $Altitude = \sqrt{(Hypothenuse)^2 - (Base)^2}.$

EXAMPLES.

1. Find the hypothenuse of a right-angled triangle whose base is 18 ft. and whose altitude is 24 ft.

SOLUTION.—Hypothenuse $= \sqrt{(18)^2 + (24)^2} = 30$ ft. *Ans.*

2. The hypothenuse of a right-angled triangle is 12½ yds. and its altitude is 10 yds. ; what is its base?

SOLUTION.—Base $= \sqrt{(12\frac{1}{2})^2 - (10)^2} = 7\frac{1}{2}$ yds. *Ans.*

3. The hypothenuse of a right-angled triangle is 7½ ft. and its base is 4½ ft. ; what is its altitude?

SOLUTION.—Altitude $= \sqrt{(7\frac{1}{2})^2 - (4\frac{1}{2})^2} = 6$ ft. *Ans.*

4. A room is 30 ft. long and 22½ ft. wide ; what is the distance between two opposite corners?   *Ans.* 37½ ft.

5. A flag-staff is perpendicular to a level plain and a rope 71¼ ft. long reaches from the top of the staff to a point of the plain 42¾ ft. from the foot of the staff; what is the height of the staff?   *Ans.* 57 ft.

6. A pair of rafters are each 22½ ft. long, and the building on which they are placed is 36 ft. wide; how high is the ridge above the plane of the eaves ?   *Ans.* 13½ ft.

7. A. and B. set out from the same point at the same time ; A. travels due north at the rate of 6 miles an hour, and B. travels due east at the rate of 4½ miles an hour; how far apart are they at the end of 3 hours ?

Ans. 22½ miles.

LENGTH OF A CIRCUMFERENCE.

283. It is shown in Geometry (B. 5, P. 11), that the circumference of a circle is equal its diameter multiplied by 3.1416; that is,

Circumference = Diameter × 3.1416.

EXAMPLES.

1. What is the length of a circumference whose diameter is 12 feet?   *Ans.* 12 *ft.* × 3.1416 = 37.6992 *ft.*

2. What is the length of a circumference whose diameter is 6.75 *ft.?*   *Ans.* 21.8058 *ft.*

3. Find the length of a circumference whose radius is 8.5 *in.*   *Ans.* 53.4072 *in.*

4. What is the circumference of a circle whose diameter is 20 yards?   *Ans.* 62.832 *yds.*

5. What is the diameter of a circle whose circumference is 78.54 *ft.?*   *Ans.* 78.54 *ft.* ÷ 3.1416 = 25 *ft.*

AREA OF A TRIANGLE.

284. It is shown in Geometry (B. 4, P. 4), that the area of a triangle is equal to half the product of its base and altitude; that is,

Area of triangle = Base × Altitude ÷ 2.

EXAMPLES.

1. The base of a triangle is 8 feet and its altitude is 6 feet; what is its area?   *Ans.* 24 *sq. ft.*

NOTE.—By the term *product of two lines* we always mean a rectangle whose length is one of the lines and whose breadth is the other. Hence we say that *the product of a line by a line is a surface.*

2. What is the area of a triangle whose base is 16 yards and whose altitude is $3\frac{1}{2}$ yards?   *Ans. 28 sq. yds.*

3. What is the area of a triangle whose base is $8\frac{1}{2}$ *yds.* and whose altitude is 14 *yds.?*   *Ans. 59½ sq. yds.*

4. The area of a triangle is 74 *sq. ft.* and its base is $9\frac{1}{4}$ *ft.;* what is its altitude?   *Ans. 16 ft.*

AREA OF A PARALLELOGRAM.

285. It is shown in Geometry (B. 4, P. 3), that the area of a parallelogram is equal to the product of its base and altitude; that is,

$$Area\ of\ parallelogram = Base \times Altitude.$$

EXAMPLES.

1. The base of a parallelogram is 14 yards and its altitude is 5 yards; what is its area?

$$Ans.\ 14\ yds. \times 5\ yds. = 70\ sq.\ yds.$$

2. Find the area of a parallelogram whose base is 13 *ft.* and whose altitude is $7\frac{1}{2}$ *ft.*   *Ans. 97½ sq. ft.*

3. A rectangle is $7\frac{1}{2}$ *rds.* long and $5\frac{1}{2}$ *rds.* wide; what is its area   *Ans. 41¼ sq. rds.*

4. A rectangular field contains 4 acres and its length is 32 rods; what is its breadth?   *Ans. 20 rds.*

AREA OF A TRAPEZOID.

286. It is shown in Geometry (B. 4, P. 5), that the area of a trapezoid is equal to the half sum of its bases multiplied by its altitude; that is,

$$Area\ of\ trapezoid = \tfrac{1}{2} \times (Upper\ base + Lower\ base) \times Altitude.$$

EXAMPLES.

1. The parallel sides of a trapezoid are 14 *yds.* and 20 *yds.* and the altitude is 7 *yds.;* what is its area?

Ans. ½ of (14 *yds.*+20 *yds.*) × 7 *yds.* = 119 *sq. yds.*

2. Find the area of a trapezoid whose parallel sides are 18 *ft.* and 22 *ft.* and whose altitude is 17 *ft.*

Ans. 340 *sq. ft.*

3. A board 14 *ft.* long is 18 *in.* wide at one end and 12 *in.* at the other end; how many square feet in its area?

Ans. 17½ *sq. ft.*

4. The parallel sides of a trapezoidal field containing 2½ acres are respectively 48 rods and 32 rods; what is the altitude, or breadth, of the field? *Ans.* 10 *rds.*

AREA OF A CIRCLE.

287. It is shown in Geometry (B. 5, P. 11), that the area of a circle is equal to the square of its radius multiplied by 3.1416; that is,

$$\textit{Area of circle} = (\textit{Radius})^2 \times 3.1416.$$

EXAMPLES.

1. What is the area of a circle whose radius is 13 inches?

Ans. 13 *in.* × 13 *in.* × 3.1416 = 530.9304 *sq. in.*

2. What is the area of a circle whose radius is 2.5 rods?

Ans. 19.635 *sq. rds.*

3. The radius of a circular fish-pond is 75 feet; what is its area? *Ans.* 17,671½ *sq. ft.*

4. The area of a circle is 176.715 *sq. ft.;* what is its radius. *Ans.* $\sqrt{176.715 \div 3.1416}$ = 7.5 *ft.*

SURFACE OF A SPHERE.

288. It is shown in Geometry (B. 8, P. 7), that the surface of a sphere is equal to 4 times the square of its radius multiplied by 3.1416; that is,

Surface of sphere $= 4 \times (Radius)^2 \times 3.1416.$

EXAMPLES.

1. What is the area of the surface of a sphere whose radius is 12 inches?

Ans. $4 \times 12 \, in. \times 12 \, in. \times 3.1416 = 1809.5616 \, sq. \, in.$

2. Find the area of the surface of a sphere whose radius is 4 feet. *Ans.* 201.0624 *sq.ft.*

3. A ten-pin ball has a surface of 78.54 *sq. in.;* what is its radius? *Ans.* $\sqrt{78.54 \, sq. \, in. \div (4 \times 3.1416)} = 2\frac{1}{2} \, in.$

4. The radius of a billiard ball is $1\frac{1}{4}$ *in.;* what is the area of its surface? *Ans.* 15.904 *sq. in.*

VOLUME, OR CONTENT, OF A PARALLELOPIPEDON, PRISM, OR CYLINDER.

289. It is shown in Geometry (B. 7, Propositions 13 and 14; B. 8, P. 1), that the content of a parallelopipedon, prism, or cylinder, is equal to the product of its base by its altitude; that is, for either of these solids we have

Content $=$ *Base* $\times$ *Altitude.*

Note.—The area of the base may often be found by one of the preceding principles.

EXAMPLES.

1. The base of a parallelopipedon is 24 *sq. ft.* and its altitude 8 *ft.;* what is its content? *Ans.* 192 *cu.ft.*

NOTE.—By the term *product of a surface and line* we always mean a parallelopipedon whose base is equal to the surface, and whose altitude is equal to the line. The number of cubic units in the volume is equal to the number of superficial units in the surface multiplied by the number of linear units in the height. Hence, we say, *the product of a surface and line,* or *the continued product of three lines, is a volume.*

2. The base of a parallelopipedon is 81 *sq. ft.* and its altitude 4 *ft.*; what is its volume?   *Ans.* 324 *cu. ft.*

3. Find the contents of a prism whose base is 86 *sq. ft.*, and whose altitude is 7 *ft.*   *Ans.* 602 *cu. ft.*

4. The base of a cylinder is equal to 80 *sq. ft.* and its altitude is equal to 5 *ft.*; what is its content?

Ans. 400 *cu. ft.*

5. The radius of the base of a cylinder is 2.5 *ft.* and its altitude is 14 *ft.*; what is its volume or contents?

Ans. (2.5 *ft.*)^2 × 3.1416 × 14 *ft.* = 274.89 *cu. ft.*

6. A stick of hewn timber is 27 *ft.* long and its cross section is 1.5 *sq. ft.*; what is its content?

Ans. 40½ *cu. ft.*

CONTENT OF A PYRAMID, OR OF A CONE.

290. It is shown in Geometry (B. 7, P. 17, and B. 8, P. 2), that the volume of a pyramid or of a cone is equal to the product of its base by ⅓ of its altitude; that is, for either of these solids we have

.   *Content = Base × Altitude × ⅓.*

EXAMPLES.

1. A base of a pyramid is 49 *sq. ft.* and its altitude is 4 *ft.*; what is its volume?

Ans. 49 *sq. ft.* × 4 *ft.* ÷ 3 = 65.3333 *cu. ft.*

2. The base of a cone is 15.9 *sq. ft.* and its altitude is 6 *ft. ;* what is its content ? *Ans.* 31.8 *cu. ft.*

3. The altitude of a cone is 18 *ft.* and the radius of its base is 4 *ft. ;* what is its content? *Ans.* 301.5936 *cu. ft.*

CONTENT OF A SPHERE.

291. It is shown in Geometry (B. 8, P. 8), that the volume or content of a sphere is equal to $\frac{4}{3}$ times the cube of the radius multiplied by 3.1416; that is,

$$\text{Content of sphere} = \frac{4}{3} \times (\text{Radius})^3 \times 3.1416.$$

EXAMPLES.

1. The radius of a sphere is 5 *ft. ;* what is its volume ?

$$\text{Ans. } \frac{4 \times 5\,ft. \times 5\,ft. \times 5\,ft. \times 3.1416.}{3} = 523.6\,cu.ft.$$

2. Find the volume of a sphere whose radius is 11.5 *ft.*
Ans. 6370.6412 *cu. ft.*

3. What is the volume of a sphere whose radius is 7½ *in.?*
Ans. 1767.15 *cu. in.*

4. The content of a sphere is 696.9116 *cu. in. ;* what is its radius? *Ans.* 5½ *in.*

BOARD MEASURE. -

292. A board foot is a solid one foot long, one foot wide, and one inch thick. It is equal to one-twelfth of a cubic foot.

This unit is used in measuring *boards, planks,* and some kinds of timber.

NOTE.—Boards and planks are of uniform thickness throughout, but they are often of different widths at the two ends ; in this case the half sum of the widths at the ends is taken as the width of the board, or plank. The width and thickness are usually expressed in inches, but the length is given in feet.

The number of board feet in a board, plank, or stick of timber may be found by the following

RULE.

Multiply the length in feet by the product of the breadth and thickness, both in inches, and divide the result by 12.

EXAMPLES.

1. How many board feet in a board 13 feet long, 16 inches wide, and $1\frac{1}{4}$ inches thick?

$$Ans. \ 13 \times 16 \times 1\tfrac{1}{4} \div 12 = 21\tfrac{2}{3}.$$

2. A board is 17 feet long, 13 inches wide at one end, 17 inches wide at the other end, and 1 inch thick; how many board feet does it contain? *Ans.* $21\frac{1}{4}$.

3. A plank 16 feet long and $2\frac{1}{4}$ inches thick is 16 inches wide at one end and 18 inches wide at the other; how many board feet does it contain? *Ans.* $56\frac{2}{3}$.

4. How many board feet in a piece of scantling $18\,ft.$ long, $4\,in.$ thick, and $9\,in.$ wide? *Ans.* 54.

TIMBER MEASURE.

293. Timber, when not measured in board measure, is usually measured in cubic feet.

Timber may be **Round**, that is, it may have a *circular cross section;* or it may be **Hewn,** that is, it may have a *rectangular cross section.*

The cross section may be the same throughout, or it may be greater at one end than at the other. The **Mean Cross Section** is the cross section midway between the two ends.

The cross section of a round stick of timber at any point can be found when we know its *girt* at that point. The **Girt** is the circumference after the bark is removed.

The cross section in square inches may be found by the

R U L E .

Multiply the square of the girt in inches by .0796.

EXAMPLES.

1. The girt of a round stick of timber is 42 inches; what is its cross section?

Ans. $42^2 \times .0796 = 140.4144$ *sq. in.*

2. Find the cross section of a round stick at a point where the girt is 52 inches.　　*Ans.* 215.2384 *sq. in.*

3. If the girt is 60 inches, what is the cross section?

Ans. 286.56 *sq. in.*

The cross section of a hewn stick of timber at any point may be found by the following

R U L E .

Multiply the breadth of the stick by its thickness at that point, both in inches; the product is the cross section in square inches.

EXAMPLES.

4. The breadth of a square stick of timber at its larger end is 14 inches and its thickness is 13 inches; what is its greatest cross section?　　*Ans.* 182 *sq. in.*

5. The breadth of the same stick at its smaller end is 12 inches and its thickness is 10 inches; what is its smallest cross section?　　*Ans.* 120 *sq. in.*

6. The breadth of the same stick at the middle of its length is 13 inches and its thickness is $11\frac{3}{4}$ inches; what is its mean cross section? *Ans.* $152\frac{3}{4}$ *sq. in.*

Knowing the two end sections and the mean section in square inches, and the length in feet, we can find the number of cubic feet in a stick of timber by the following

RULE.

To the sum of the end sections add four times the mean section, all in square inches, and multiply the result by the length in feet; then divide by 864, and the quotient will be the number of cubic feet required.

EXAMPLES.

7. The end sections of a stick of timber are 182 and 120 *sq. in.*, the middle section is $152\frac{3}{4}$ *sq. in.*, and its length is 30 *ft.*; what is its content?

$$Ans. \quad \frac{(182 + 120 + 4 \times 152\frac{3}{4}) \times 30}{864} = 31.7014 \, cu.ft.$$

8. The end sections of a stick of timber 40 *ft.* long are 460 and 400 *sq. in.*, and the mean section is 440 *sq. in.*; what is its content? *Ans.* 121.2963 *cu. ft.*

METHOD OF DUODECIMALS.

294. If the linear dimensions are expressed in feet and inches, areas and volumes may be found by the **Method of Duodecimals.**

In this system of numbers the primary unit is 1 *foot;* it may be a *linear foot*, a *square foot*, or a *cubic foot.*

One twelfth of a foot is called a **Prime,** one twelfth of a prime is called a **Second,** and one twelfth of a second is called a **Third,** as shown in the following

TABLE.

12 thirds ‴ make 1 second..... ″.

12 seconds " 1 prime...... ′.

12 primes " 1 foot........ $\begin{cases} ft. \\ sq.\,ft. \\ cu.\,ft. \end{cases}$

The scale of the system is uniform, that is, it is

12,　　12,　　12.

In accordance with the principles laid down for multiplying lines by lines, and surfaces by lines, we see that

feet	multiplied by	*feet*	give	*feet ;*
feet	"	" *primes*	"	*primes ;*
feet	"	" *seconds*	"	*seconds ;*
primes	"	" *primes*	"	*seconds ;*
primes	"	" *seconds*	"	*thirds.*

OPERATION OF MULTIPLICATION.

295. Let it be required to find the continued product of 3 *ft.* 5 *in.*, 2 *ft.* 6 *in.*, and 4 *ft.* 7 *in.*:

EXPLANATION.—Having written the first two numbers so that units of the same name stand in the same column, we begin at the left hand and multiply all the parts of the multiplicand by 2, writing the products, without reduction, in their proper columns according to the principles explained in the last article. We then multiply all the parts of the multiplicand by 6, and place the products in their proper columns, as determined by the rules in the last article. We next add the partial products by the rule for addition of compound numbers, which gives 8 *sq. ft.* 6′ 6″.

OPERATION.

ft.	′	″	‴
3	5		
2	6.		
6	10		
	18	30	
8	6	6	
4	7		
32	24	24	
	56	42	42
39	1	9	6

$$= 39\tfrac{43}{288} \ cu.ft.$$

We now multiply 8 *sq. ft. 6' 6''* by 4 *ft. 7 in.*, in the same manner as before and find for the required product 39 *cu. ft. 1' 9'' 6'''*, that is $(39 + \frac{1}{12} + \frac{9}{144} + \frac{6}{1728})$, *cu. ft.*, which is equal to $39\frac{48}{288}$ *cu. ft.*

In like manner we may multiply in all similar cases; hence, the following

RULE.

I. Write the numbers so that units of the same name shall stand in the same column.

II. Multiply all the parts of the multiplicand by each part of the multiplier and write the corresponding partial products, without reduction, in their proper columns.

III. Add the partial products by the rule for addition of compound numbers.

EXAMPLES.

	(1.)			(2.)		
	ft.	*'*	*''*	*ft.*	*'*	*''*
Multiplicand....	3	2		5	7	
Multiplier	5	7		7	10	
	15	10		35	49	
		21	14		50	70
Product.......	17	8	2	43	8	10
	$= 17\frac{48}{72}$ *sq. ft.*			$= 43\frac{43}{72}$ *sq. ft.*		

3. Multiply 3 *ft.* 7 *in.* by 9 *ft.* 4 *in.*

Ans. 33 *sq. ft.* 5' 4'' = $33\frac{4}{9}$ *sq. ft.*

4. Multiply 5 *ft.* 11 *in.* by 16 *ft.* 2 *in.*

Ans. 95 *sq. ft.* 7' 10'' = $95\frac{43}{72}$ *sq. ft.*

5. Find the continued product of 3 *ft.* 4 *in.*, 2 *ft.* 7 *in.*, and 6 *ft.* 11 *in.* *Ans.* 59 *cu. ft.* 6' 8'' 8''' = $59\frac{131}{216}$ *cu. ft.*

6. Find the continued product of $4\,ft.\ 3\,in.$, $5\,ft.\ 2\,in.$, and $6\,ft.\ 5\,in.$ $Ans.\ 140\,cu.\,ft.\ 10'\ 9''\ 6''' = 140\frac{2\,6\,4}{8\,6\,4}\,cu.\,ft.$

PRACTICAL PROBLEMS.

1. How many square feet in a ceiling $17\,ft.\ 3\,in.$ long and $11\,ft.\ 5\,in.$ wide? $Ans.\ 196\frac{11}{16}\,sq.\,ft.$

2. How many square feet in a pavement $12\,ft.\ 6\,in.$ long and $10\,ft.\ 2\,in.$ wide? $Ans.\ 127\frac{1}{12}\,sq.\,ft.$

3. Find the capacity of a box $3\,ft.\ 3\,in.$ long, $2\,ft.\ 9\,in.$ wide, and $1\,ft.\ 11\,in.$ deep. $Ans.\ 17\frac{25}{192}\,cu.\,ft.$

4. Find the contents of a stick of timber $42\,ft.\ 6\,in.$ long, $1\,ft.\ 7\,in.$ wide, and $1\,ft.\ 4\,in.$ thick.

$Ans.\ 89\frac{13}{18}\,cu.\,ft.$

5. What is the capacity of a bin $7\,ft.\ 3\,in.$ long, $4\,ft.\ 2\,in.$ wide, and $3\,ft.\ 5\,in.$ deep? $Ans.\ 103\frac{61}{288}\,cu.\,ft.$

6. How many cords of wood in a pile $13\,ft.\ 3\,in.$ long, $4\,ft.\ 2\,in.$ wide, and $3\,ft.\ 6\,in.$ high?

$Ans.\ 1\ C.\ 65\frac{11}{48}\,cu.\,ft.$

7. How many cords in a pile $20\,ft.\ 4\,in.$ long, $4\,ft.\ 3\,in.$ wide, and $5\,ft.\ 2\,in.$ high? $Ans.\ 3\ C.\ 62\frac{35}{72}\,cu.\,ft.$

8. How many cubic feet of stone in a wall $27\,ft.\ 6\,in.$ long, $3\,ft.\ 3\,in.$ thick, and $4\,ft.\ 2\,in.$ high?

$Ans.\ 372\frac{13}{48}\,cu.\,ft.$

9. What is the area of a rectangle whose length is $9\,ft.\ 7\,in.$, and whose breadth is $7\,ft.\ 4\,in.$?

$Ans.\ 70\frac{5}{18}\,sq.\,ft.$

10. The base of a cylinder is $24\,sq.\ in.$ and its altitude is $2\,ft.\ 9\,in.$; what is its content? $Ans.\ 66\,cu.\,ft.$

11. What is the content of a room whose length is $18\,ft.$

6 *in.*, whose breadth is 12 *ft.* 4 *in.*, and whose height is 10 *ft.* 2 *in.?*   *Ans.* 2,319$\frac{24}{36}$ *cu. ft.*

12. What is the area of a floor whose length is 25 *ft.* 3 *in.*, and whose breadth is 20 *ft.* 6 *in.?*

Ans. 517$\frac{5}{8}$ *sq. ft.*

13. What is the content of a box 1 *ft.* 6 *in.* long, 1 *ft.* 3 *in.* wide, and 1 *ft.* 1 *in.* deep ?   *Ans.* 2$\frac{1}{32}$ *cu. ft.*

14. What is the content of a cube, each edge of which is 3 *ft.* 4 *in.?*   *Ans.* 37$\frac{1}{27}$ *cu. ft.*

REVIEW QUESTIONS.

(**271.**) What is mensuration? (**272.**) What is a polygon? Sides? Vertices? (**273.**) What is the area of a polygon? (**274.**) Define a triangle. Its base. Its altitude. (**275.**) What is a parallelogram? A rectangle? (**276.**) What is a trapezoid? Its lower and its upper bases? Its altitude? (**277.**) What is a prism? Its bases? Its lateral faces? Its altitude? (**278.**) What is a pyramid? Its base? Its lateral faces? Its altitude? What is a frustum of a pyramid? (**279.**) What is a cylinder? Its bases? Its convex surface? Its altitude? (**280.**) What is a cone? Its base? Its convex surface? Its vertex? Its altitude? What is a frustum of a cone? (**281.**) What is a sphere? Its centre? A diameter? A radius? (**282.**) What is the relation between the sides of a right-angled triangle? (**283.**) What is the length of a circumference? (**284.**) The area of a triangle? (**285.**) Of a parallelogram? (**286.**) Of a trapezoid? (**287.**) Of a circle? (**288.**) Of the surface of a sphere? (**289.**) What is the content of a parallelopipedon, prism, or cylinder? (**290.**) Of a pyramid or cone? (**291.**) Of a sphere? (**292.**) What is a board foot? How do you find the number of board feet in a board or plank? (**293.**) How is timber measured? (**294.**) What is the method of duodecimals? Define primes, seconds, and thirds. (**295.**) Give the rule for multiplication.

Miscellaneous Examples.

1. Find the product of the sum and difference of 25 and 16.

2. Divide the difference between 1296 and 441 by the sum of 36 and 21.

3. What are the prime factors of 9,800?

4. Resolve 3,990 into prime factors?

5. Find the *g. c. d.* of 2,290 and 458.

6. What is the *g. c. d.* of 1,435, 1,085, and 2,135?

7. Find the *l. c. m.* of 15, 18, 24, 40, and 50.

8. What is the *l. c. m.* of 508 and 889?

9. Add $\frac{2}{3}$, $\frac{1}{7}$ of $\frac{4}{7}$, and $3\frac{4}{5}$.

10. Subtract $\frac{1}{3}$ of $4\frac{1}{2}$ from $\frac{6}{7}$ of $9\frac{1}{4}$.

11. Multiply $\frac{1}{4}$ of $2\frac{1}{4}$ by $\frac{1}{8}$ of $3\frac{1}{7}$.

12. Divide $2\frac{1}{2}$ by $1\frac{4}{5}$.

13. Multiply $3.31 + 4.06$ by $8.13 - 3.43$.

14. Divide $3.8 + 2.05$ by $8.6 - 3\frac{3}{4}$.

15. A man bought a horse and carriage; the horse cost $\frac{3}{5}$ as much as the carriage, and both together cost $640; what was the cost of the horse?

16. At a certain election the successful candidate had a majority of 120, which was $\frac{1}{4}$ of all the votes cast; how many votes did the opposing candidate receive?

17. Divide \$357 among A., B., and C., so that B. shall receive 2½ times as much as A., and C. as much as A. and B. together.

18. A. can do a piece of work in 3 days, B. can do it in 4 days, and C. can do it in 5 days; how long will it take them to do it together?

19. How many bushels of oats can be raised on 4½ acres, if each acre produces 47 *bu.* 3 *pks.*?

20. How many bushels of wheat at \$1.75 per bushel will it take to pay for 3 *cwt.* of pork at \$7 per *cwt.*?

21. A grocer mixes 120 *lbs.* of sugar at 10 *cts.* a pound, 140 *lbs.* at 12 *cts.*, and 60 *lbs.* at 14 *cts.*; at what rate must he sell it to clear 20% on its cost?

22. Divide \$1,000 among 3 persons in the proportion of 5, 7, and 8.

23. Bought a horse for \$312 and sold him at a loss of 12½%; what did I receive?

24. A man travels 100 miles by rail and 100 miles by stage; his average rate of travel is 16 miles per hour and the rate of the train is 40 miles per hour; what is the rate of the stage?

25. A tank 7 *ft.* deep, 12 *ft.* long, and 9 *ft.* broad is full of water; what is the weight of the water, if each cubic foot weighs 62½ *lbs.*?

26. Three men can do a piece of work in 6 days; the first can do it in 15 days, and the second can do it in 12 days; how long would it take the third to do it?

27. A man bequeathed \$37,000 to his family; he gave ¼ to his wife, ⅓ to his son, and divided the rest equally between 5 daughters; how much did each daughter receive?

28. A merchant purchased cloth to the amount of $1,250 and silk goods to the amount of $900 ; he sold the former at a profit of 20% and the latter at a loss of 10%; how much did he gain by the operation ?

29. A., B., and C. enter into partnership; A. puts in ½ of the capital, B. puts in ¼ of the capital, and C. puts in the rest; at the end of the year their profit amounts to $10,440; what is C.'s share of the profit ?

30. I bought a lot for $700 and after holding it for 1 year, sold it at an advance of 20% on the cost and interest at 7%; what did I get?

31. A merchant bought cloth to the amount of $1,500 and sold it again for $1,770; what was his gain %?

32. A merchant bought goods for $3,000 cash, and sold them again for $3,810 on a credit of four months; what was his gain % in addition to interest on his money at the rate of 6%?

33. A merchant sells cloth at $3.12½ per yard and clears 25%; what % would he clear if he were to sell it for $3.50 per yard ?

34. Two couriers start together from the same point and travel in the same direction; the first travels 23 miles in 3½ hours and the second travels 11 miles in 2¼ hours; how far apart are they at the end of 31½ hours ?

35. A laborer spent 25% of his week's wages for flour and had $11.25 left; what did he receive per week ?

36. John Churchill's farm is composed of 36 acres of pasture land, 22 of meadow land, 18 of plough land, and 20 of woodland; supposing it to be rectangular in shape and 128 rods long, what is its width ?

37. If 30 bushels of wheat cost \$67.50, how much can be bought for \$438.75 ?

38. If 12 men can build a wall in 20 days, how many men will be required to do the same work in 8 days ?

39. If \$100 gain \$6 in 9 months, what principal will gain \$11 in 5 months ?

40. A certain quantity of hay will feed 963 sheep for 7 weeks; how many must be turned away that it may feed the remainder for 9 weeks ?

41. The third part of an army were killed, the fourth part were taken prisoners, and there remained 10,800; how many men did the army contain ?

42. Thomas sold 600 pineapples at $16\frac{2}{3}$ cts. each, and received as much as Henry did for a number of melons at 40 cts. ; how many melons did Henry sell ?

43. A flag-staff stands $\frac{1}{4}$ of its length in the ground, 12 feet in the water, and $\frac{4}{8}$ of its length in the air ; what is the length of the staff ?

44. S., J., and B. enter into partnership ; S. puts in \$5,600, J. \$4,900, and B. \$3,500; if they gain \$1,650, how much will each gain ?

45. A shipper insures \$2,500 worth of oats at 4% ; for what must he insure to receive the value of his oats and the cost of insurance in case of loss ?

46. A merchant insures \$4,000 worth of silk at $2\frac{1}{2}$% ; what must be the face of his insurance that he may lose nothing in case of its destruction by fire ?

47. A merchant bought several bales of cloth, each containing $133\frac{1}{3}$ yds., at the rate of 12 yds. for \$11, and sold

it at the rate of 8 *yds.* for $7, losing $100 by the transaction ; how many bales did he buy ?

48. A. owes B. $2,500, payable in 4 months, but at the end of 3 months he pays him $1,500 ; how long after this payment before the balance is equitably due ?

49. What is the bank discount on an accommodation note of $3,000 for 60 days at 7% ?

50. A lady wishes to carpet a floor 15 *ft.* 9 *in.* wide and 22 *ft.* 6 *in.* long, with carpeting ¾ *yd.* wide ; if the carpeting is worth $2.50 per *yd.* how much will it cost ?

51. Three men hire a pasture for 1 year and pay $45 for its use ; the first puts in 100 head of cattle, the second puts in 150 head, and the third puts in 50 head : what must each pay ?

52. How many board feet in 250 planks, each 14 *ft.* long, 16 *in.* wide, and 2½ *in.* thick ?

53. A pile of wood is 3¼ feet wide, 5¼ feet high, and 147 feet long ; how many cords does it contain ?

54. A., B., and C. undertake a job for $400 ; A. furnishes 4 men for 8 days, B. 6 men for 7 days, and C 13 men for 2 days : what share of the money ought each to receive ?

55. A merchant bought a piece of merino containing 32 yards for $25.60, and then marked it so that he could fall 4% on the asking price and still make 20% on its cost ; what did he mark it per yard ?

56. A house is 40 feet from the ground to the eaves, and a ladder is placed with its foot 30 feet from the house ; how long must it be to reach the eaves ?

57. A man buys ⅝ of a piece of property and sells 20%

of his share for $5,000, clearing 25% on its cost; what was the original value of the whole property?

58. If wheat at 8*s*. 3*d*. per bushel gives a profit of 10%, how much will it give if sold at 9*s*. 4½*d*. per bushel?

59. Of the trees in an orchard ½ are apple trees, ¼ peach trees, ⅛ plum trees, and the remaining 200 are cherry trees; how many trees in the orchard?

60. If a quantity of bread will last 1,500 men for 12 weeks at the rate of 20 *oz*. per day for each man, how long will the same bread last 2,500 men at the rate of 16 *oz*. per day for each man?

61. A path 3 feet wide runs around a rectangular yard whose length is 105 *yds*. and whose breadth is 95 *yds*.; if the outer edge of the walk is 4 feet from the wall, how many square feet will it contain?

62. A man leaves $38,000 to be divided among 3 sons and 3 daughters; each son is to receive 33⅓% more than the eldest daughter and each of the younger daughters is to receive 33⅓% less than the eldest: what is the share of each?

63. If a clock beats 31 times in 30 seconds, how many times will it beat in 3 *da*. 5 *hrs*. 4 *min.?*

64. A man agrees to execute a contract in 60 days and places 30 men on the work; at the end of 48 days the job is but half completed: how many men must he employ the rest of the time to fulfill his contract?

65. A man sells eggs; to the first person he sells half his stock and *one* more, to the second person he sells half of what remains and *one* more, and to the third person he sells half of what remains and *one* more, when he has none left: how many had he at first?

66. At what time between 5 and 6 o'clock are the hour and minute hands of a clock together ?

67. If 30 men require 40 days to do a piece of work, how many men will be required to do 5 times as much work in one fifth of the time ?

68. A gentleman being asked his age, replied, if you add to it its half, its third, and three times three, the sum will be 130 ; what was his age ?

69. A. and B. together can do a piece of work in 18 days, but with the assistance of C. they can do it in 11 days ; in what time could C. do it by himself ?

70. A. starts from Bantam at 9 h. 15 m. A. M. and travels toward Norfolk at the rate of 4 miles per hour ; B. starts from Norfolk at 9 h. 30 m. A. M. and travels toward Bantam at the rate of $3\frac{1}{2}$ miles per hour ; the distance between the two places is 21 miles : at what time will they meet ?

71. If 3,000 copies of a book of 11 sheets require 66 reams of paper, how much paper will 5,000 copies of a book of $12\frac{1}{2}$ sheets require ?

72. If 24 men can reap 76 acres in 6 days, how long will it take 18 men to reap 114 acres ?

73. If 10 men can blast 30 $cu.\ yds.$ of rock in 8 days, how many $cu.\ yds.$ can 20 men blast in 10 days ?

74. If 7 men can mow 84 acres in 12 days of $8\frac{1}{4}$ hours each, in how many days of $7\frac{1}{2}$ hours each can 20 men mow 208 acres ?

75. How many acres of land, at $150 per acre, must be given for 750 $bbls.$ of flour, at $4.60 per barrel ?

76. How many kilogrammes of butter, at 50 $cts.$ a

kilog., must be given for 7 *meters* of cloth, at $4.50 per meter ?

77. How much cloth, at 20 *fr.* per meter, must be given for a watch worth 315.20 *fr.?*

78. If 3 trees furnish 8.1 *steres* of timber, what will 62 trees of the same kind produce ?

79. If 5 children eat 21 *hectog.* of cake, how much will 93 children eat ?

80. Divide 55 *fr.* 35 *c.* among 7 men and 6 women, giving to each man 3 times as much as to each woman.

81. A. and B. commence trade ; A. puts in $350 for 8 months, B. $600 for 7 months, and they make $700 ; to what part of the gain is each entitled ?

82. W. and B. engage in business ; W. puts in $18,000 for 17 *mo.* and B. puts in $24,000 for 6 *mo. ;* while in business they lose $6,500 ; what loss must each bear?

83. On the 1st of January, A. commenced business with a capital of $17,000 ; on the 1st of April B. entered the business, advancing $12,000 capital ; and on the 1st of July C. was admitted and advanced $16,000 ; at the end of the year the firm had gained $8,160. How much of the gain ought each to receive ?

84. A., B., and C. have business transactions together whereby they gain $18,049.60 ; A. furnished $22,000 for 12 months, B. $18,600 for 10 months, and C. $30,000 for 7 months ; to what part of the gain is each entitled?

85. A grocer has two kinds of tea ; the better kind is worth $1.20 a pound and the poorer kind is worth 75 *cts.* a pound ; in what proportion must he mix them that the mixture may be worth $1 a pound ?

86. A. and B. together can do a job in 7 days, but it would take A. alone 12 days to do it; how long would it take B. alone to do it?

87. A father left $10,000 to his two sons, aged respectively 14 and 18, to be divided between them so that the shares at simple interest at 5% should be the same when each was 21 years old; what was the share of each?

88. A. and B. commenced business with equal sums of money; at the end of one year A. had gained a sum equal to $\frac{1}{3}$ of his original capital and B. had lost $5,000; A. then had twice as much as B.; what was the original capital of each?

89. A man divided his estate into three equal parts, giving to his wife $200 more than $\frac{1}{3}$ of the whole; to his son $400 more than $\frac{1}{3}$ of the whole, and to his daughter $600 more than $\frac{1}{3}$ of the whole; what was the value of the estate?

90. A manufacturer employed men, women, and boys; he had 3 women to every 2 men, and 3 boys to every 2 women; to the men he paid $1, to the women 50 *cts.*, and to the boys 25 *cts.* a day; at the end of 6 days he paid them all $222. How many men did he employ?

91. The head of a fish was 9 inches long; its tail was as long as its head and half of its body; and its body was as long as its head and tail together; how long was the fish?

92. A father distributed to his three sons A., B., and C., a sum of money, giving to A. $4 as often as to B. $3, and to C. $5 as often as to B. $6; if A.'s share is $5,000, what does each of the others get?

93. A prize of $945 is divided amongst a captain, 4 men,

and 1 boy; the captain has 1½ shares, each man 1 share, and the boy ⅓ of a share; what does each receive ?

94. A bankrupt's assets amounted to $4,000; to A. he owed ½ of the assets, to B. ⅓ of the assets, to C. ¼ of the assets, and to D. ⅕ of the assets. The entire assets being divided among these 4, what did A. receive?

95. How many posts 7 *ft.* apart will be required in fencing a rectangular lot containing 70,756 *sq. ft.*, the length of the lot being 4 times its breadth ?

96. Divide $1,000 amongst A., B., and C., so that A. shall have $100 more than C., and B. $95 less than C.

97. What number is that from which if you take ⅚ of ⅔ and to the remainder add $\frac{7}{16}$ of $\frac{1}{20}$, the result will be 10 ?

98. A person asked the hour of the day and was told that the time past noon was ⅘ of the time to midnight; what was the time?

99. When a man was married he was 3 times as old as his wife, but 15 years afterward he was only twice as old; what was his age when he was married ?

100. A man going to market was met by another, who said, "Good morrow, with your 100 geese." He replied, "I have not a hundred geese, but if I had half as many more as I have, and 2½ geese more, I should have a hundred;" how many geese had he ?

ANSWERS.

9. 14,776.
10. 11,803.
11. 52,026.
12. 11,594 *ft.*
13. 151,275.
14. 7,618 *yds.*
15. 16,082 *da.*
16. 1,038,957.
17. 7,290.
18. 19,585.
21. 82,391.
22. 779,264 *yds.*
23. $1,624,249.
24. $686,853.
25. 273,329.
26. 253,693 *ft.*
27. $1,041,262.
28. 564,407.
29. $351,405.
30. 206,317.
31. 13,507.
32. $839.
33. 179,580 *yds.*
34. 1,715,099.
35. 1,715,369.
35. 14,759,180.
37. 2,159,170.
38. 1,707,521.
39. 6,982,126.
40. 12,433,713.
41. 7,921,317.
42. 94,370,040.
43. $953.94.
44. $1,688.54.
45. $930.23.
46. $2,808.70.
47. $106,059.69.

48. $2,413,450.43.
49. $291,146,187.88.
50. $5,822.08.

Problems.

8. 254.
9. $16,985.
10. 107.683 *bu.*
11. 587 *mi.*
12. 258,928 *ft.*
13. 3,695 *pp.*
14. $92,950.
15. 3,073,134.
16. 74,470 *mi.*
17. 44,437,245.
18. 48,154 *yds.*
19. 37,199.
20. 39,351 *mi.*
21. $75,063.22.
22. $925.90.
23. $59,984.48.
24. $1,136.98.

13. 17,571.
14. 18,654.
15. 23,017.
16. 57,921.
17. 19,238.
18. 591,203.
19. 666,667.
20. 78,004.
21. 900,497.
22. 305,106.
23. 37,486.
24. 111,530.
25. 409,095.
26. 561,906,000.
27. 604,918.

28. 19,712.
29. 402,760.
30. 39,990,990.
31. 61,303.
32. 78,325.
33. 5,322.
34. 28,571.
35. $722,996.
36. $206,992.
37. $801,965.
38. 33,522 *yds.*
39. $57,838,447.
40. $4,312,956.
41. 2,911,106.
42. 178,514.
43. 39,499.
44. 4,880,874.
45. 1,815,309.
46. 8,600,090.
47. 959,820 *ft.*
48. 481,605.
49. 520,619.
50. 10,875.
51. 17,138.
52. 33,335.
53. $33,335.
54. 74,825 *ft.*
55. 345,153.
56. $24.51.
57. $1,781.13.
58. $21,186.73.
59. $4,349.66.
60. $75,785.11.

Problems.

5. 3,602.
6. 71,837.
7. 30,388.
8. 38,788.
9. $52,806.

10. $36,861.
11. $397.
12. $883.
13. $2,004.
14. $577.
15. 9,417 *A.*
16. $25,600.
17. $9,112.
18. $6,740.
19. 1897.
20. 28,887 *bu.*
21. *Lost* $2,410.
22. 31 *mi.*
23. 123 *mi.*
24. $4,419.84.
25. $3,853.48.
26. 2,331 *sq. mi.*
27, $351.74.
28. 1,620.
29. $5,318.
30. 10,680.

Art. 29.
Page 40.

9. 677,184.
10. 203,940 *ft.*
11. 244,494 *lbs.*
12. $1,665.255.
13. 394,875.
14. 163,636 *ft.*
15. $1,227,142.
16. 957,504.
17. 284,733.
18. 1,039,668.
19. $717,552.
20. $4,233,086.
21. 5,737,401.
22. 5,333,328.

Art. 32.
Pages 43–44.

8. 305.375.
9. $398,088.
10. 131,794.
11. 76,923.
12. 536,724.
13. 337,770.
14. 2,032,128.
15. 3,129,385.
16. 2,807,208.

17. 5,760,757.
18. 38,801,217.
19. 16,179,212.
20. 45,656,744.
21. 183,280,678.
22. 86,409,776.
23. 129,414.654.
24. 110,083,096.
25. 28,370,748.
26. 143,533,733.
27. 437,557,351.
33. 6,786,000.
34. 27,318,000.
35. 33,948,000.
36. 15,400,800.
37. 516,672,000.
38. 1,560,793,500.
39. 64,090,000.
40. 16,442,400.
41. 9,396,000.
42. 304,741,000.
43. 3,179,520,000.
44. 369,369,000.
45. 1,049,760,000.
46. 300,000,000
47. 3,458,280,000.
48. 903,243,000.
49. 183,293,000,000.
50. 33,442,200,000.
51. 3,199,878.
52. 52,970,405.
53. 13,642,498.
54. 47,673,087.
55. 630,063,000.
56. 1,100.220,680.
57. 169,589.100.
58. 2,716,002.
59. 120,051.
60. 11,875,160.

Art. 34.
Page 45.

2. 125,712.
3. 74,508 *ft.*
4. $242,235.
5. 3,672,672 *yds.*
6. 1,034,352 *lbs.*
7. 35,328.
8. 759,440.
9. $1,820,808.

10. 126.720.
11. 275,184.

Problems.
Pages 45–48.

5. 119,568 *min.*
6. 262,800 *bbls.*
7. 16,000 *rds.*
8. $16,362,500.
9. 1,624 *bu.*
10. 4,111,200 *ft.*
11. 17,920 *rds.*
12. 51,574.
13. 153,180.
14. 349,860 *lbs.*
15. 95,040 *ft.*
16. 40,824.
17. $202.50.
18. $11.58.
19. $11,400.
20. 30 *mi.*
21. 854,496.
22. 172 *mi.*
23. 223 *mi.*
24. 1,240 *mi.*
25. 792 *yds.*

Art. 40.
Page 54.

15. 3,090.
16. 746.
17. 3,367.
18. 9,476.
19. 11,359.
20. 91,477.
21. $91,306$\frac{2}{3}$.
22. 57,799$\frac{1}{4}$.
23. 45,902 *lbs.*
24. 143,071$\frac{1}{4}$.
25. $2,379,590$\frac{1}{4}$.
26. 10,290,589$\frac{5}{8}$ *yds.*
31. 74,074.
32. 7,007.
33. 619.
34. 2,228.
35. 12,902.
36. 1,344.
37. 11,451$\frac{6}{12}$.
38. 10,351.
39. 5,972.

40. $1,519$\frac{7}{18}$.
41. 2,216$\frac{4}{15}$.
42. 33,600$\frac{10}{12}$.
43. $1,006$\frac{2}{3}$.
44. 561$\frac{2}{3}$.
45. $3.11.
46. 2\frac{4}{16}$.
47. 415 *lbs.*
48. $9.21.
49. $8.72.
50. 468 *yds.*
51. $4.18.
52. 67 *ft.*

Art. 41.
Page 57.

5. 217.
6. 342.
7. 226.
8. 105$\frac{123}{113}$.
9. 102.
10. 99.
11. 461.
12. 72.
13. 284.
14. 481$\frac{80}{154}$.
15. 217.
16. 218$\frac{17}{209}$.
17. 463$\frac{11}{112}$.
18. 1,003.
19. 7,815.
20. 96.
21. 192 *lbs.*
22. 85 *yds.*
23. 1,766$\frac{2}{3}$ *ft.*
24. $356.
25. 34.
26. 201.
27. $37.
28. 195.
29. 2,503 *mi.*
30. 203 *horses.*
31. 7,941.
32. 3,864.
33. 2,372$\frac{30}{73}$.
34. 133,056.
35. 165,503.
36. 844,101.
37. 34,807.
38. 1,684.

39. 14,076.
40. 997.
41. $32.83.
42. 17,544$\frac{11}{54}$.
43. 1,345.
44. 194,877$\frac{49}{74}$.
45. $32,528.17.
46. $325.49.
47. 2,017$\frac{108}{216}$.
48. 538$\frac{46}{85}$.
49. $70.56.
50. $672.90.
51. 444.
52. 71.
53. 37.
54. 869.
55. 17.

Art. 42.
Pages 58-60.

4. 86.
5. 45,561$\frac{15}{42}$.
6. 213.
7. 864.
8. 6,129$\frac{11}{110}$.
9. 13,922$\frac{60}{105}$.
10. 245$\frac{14}{110}$.
11. 405$\frac{141}{231}$.
13. 74$\frac{6}{10}$.
14. 13$\frac{88}{100}$.
15. 49$\frac{81}{100}$.
16. 8$\frac{687}{1000}$.
17. 34$\frac{25}{100}$.
18. 9,427$\frac{6}{10}$.
21. 4$\frac{88}{99}$.
22. 3$\frac{1088}{2500}$.
23. 10$\frac{489}{1700}$.
24. 30$\frac{14810}{41000}$.
25. 30$\frac{1896}{4200}$.
26. 20$\frac{30741}{84000}$.
28. 7,630.
29. $370.
30. 25.
31. 56.
32. 77.
33. 825$\frac{45}{47}$.
34. 4,860.
35. 11$\frac{881}{...}$.
36. 7$\frac{88}{880}$.
37. 34,492.

38. 5$\frac{432}{860}$.
39. 7$\frac{16}{48}$.
40. 5$\frac{5}{27}$.
41. 87$\frac{24}{42}$.

Art. 43.
Problems.
Pages 61-64.

6. $118.
7. 448 *A.*
8. 123 *rows.*
9. 108.
10. 284 *bbls.*
11. 42 *yds.*
12. 92 *A.*
13. 2,727 *yds.*
14. $247.
15. $122.
16. $1,998.
17. $10,405.
18. 23,709.
19. 560 *mi.*
20. $1,746.
21. 11.749 *lbs.*
22. 12,768 *yds.*
23. 815.
24. 140 *yds.*
25. $29,650.
26. $25,175.
27. $12.
28. 36 *hrs.*
29. 2,074 and 8,296.
30. 312.
31. 1000.
32. *dau.* $12,923 ; each son $13,763.
33. $185.25.
34. 102 *da.*
35. 132.
36. 48 *bbls.*
37. 114 *mi.*
38. $23.
39. $86 *A,* at $42 an *A.*
40. 66 *horses.*

Art. 48.
Pages 67-68.

8. 2. 3. 67.
9. 3. 7. 79.
10. 2. 5. 7. 47.

11. 2. 3. 7. 37.	**Art. 54.**	3. 1,008.	9. $\frac{247}{16}$.
12. 3. 5. 73.	*Page 74.*	4. 5,048.	10. $\frac{151}{19}$.
13. 2. 3. 5. 7. 11.	2. 4.	5. 13,860.	11. $\$\frac{191}{10}$.
14. 2. 3. 5. 7. 13.	3. 45.	6. 2,520.	12. $1\frac{18}{?}$ *lbs.*
15. 2. 2. 2. 3. 5. 11. 13.	4. 63.	7. 540.	13. $1\frac{7}{9}$ *yds.*
16. 2. 2. 3. 7. 7. 13.	5. 15.	8. 420.	14. $7\frac{85}{9}$ *hrs.*
17. 2. 2. 2. 2. 3. 3. 13.	6. 66.	9. 720.	15. $\frac{353}{18}$.
18. 2. 3. 131.	7. 210.	10. 1,176.	16. $\frac{258}{17}$.
19. 2. 2. 2. 2. 2. 2. 7. 7.	8. 64.	11. 16.800.	17. $\frac{2605}{59}$.
20. 2. 3. 5. 31.	9. 108.	12. 3,528.	18. $\frac{784}{25}$.
21. 3. 5. 97.	10. 81.	13. 44,100.	19. $\frac{359}{16}$.
22. 5. 11. 67.	11. 42.	14. 14,700.	20. $\frac{1007}{20}$.
23. 2. 11. 79.	12. 630.	15. 468.	21. $\frac{1022}{10}$.
24. 3. 5. 7. 31.	13. 267.	16. 5,070.	22. $\frac{278}{11}$.
25. 7. 11. 13.	14. 396.	17. 3,400.	23. $\frac{212}{9}$.
26. 5. 13. 17.	15. 13	18. 4,275.	24. $\frac{1545}{18}$.
27. 3. 7. 151.		19. 13,475.	25. $\$\frac{959}{?}$.
28. 13. 13. 17.	**Art. 56.**	20. 1,512.	26. $\$\frac{950}{?}$.
	Pages 75-76.		27. $\$\frac{1051}{?}$.
Art. 51.	1. 267.	**Problems.**	28. $\frac{848}{4}$ *lbs.*
Page 70.	2. 396.	*Page 79.*	29. $\frac{2900}{?}$.
7. 16.	3. 414.	1. 9 or $9.	30. $\frac{6018}{?}$.
8. 93.	4. 532.	2. 180 *ft.*	31. $\frac{14481}{37}$.
9. 52.	5. 84.	3. 45 *bu.*	32. $\frac{43627}{?}$.
10. $41\frac{6}{35}$.	6. 33.	4. 120 *ft.*	33. $\frac{8857}{?}$.
11. 400.	7. 23.	5. 12.	34. $\frac{18585}{99}$.
12. 57.	8. 87.	6. 30 *qts.*	35. $\frac{21192}{18}$.
13. 21.	9. 630.		36. $\frac{226979}{598}$.
14. $14\frac{7}{22}$.	10. 267.	**Art. 67.**	37. $\frac{410871}{98}$.
15. $5\frac{5}{14}$.	11. 396.	*Page 85.*	38. $\frac{317821}{?}$.
16. $26\frac{8}{37}$.	12. 72.	1. $\frac{48}{?}$.	39. $\frac{6878}{8}$ *lbs.*
17. 5.	13. 135.	2. $\frac{70}{5}$.	40. $\frac{5196}{?}$.
18. 1.	14. 72.	3. $\frac{21}{4}$.	41. $\frac{1461}{4}$ *da.*
19. $1\frac{6}{23}$.	15. 23.	4. $\frac{9.5}{?}$.	42. $\$\frac{8484}{11}$.
20. $1\frac{1}{3}$.	16. 252.	5. $\frac{462}{11}$.	
21. $15\frac{3}{34}$.	17. 3.	6. $\frac{2085}{5}$.	**Art. 69.**
22. 3.	18. 12.	7. $\frac{6148}{8}$.	*Pages 87, 88.*
23. 2.	19. 8.	8. $\frac{8908}{8}$.	1. 125.
24. $6\frac{1}{2}$.	20. 4.	9. $\frac{27884}{84}$.	2. 125.
	21. 37.	10. $\frac{41892}{57}$.	3. $36\frac{1}{2}$.
Problems.	22. 108.		4. $31\frac{?}{?}$.
5. 28 *cts.*	23. 73.	**Art. 68.**	5. $31\frac{7}{?}$.
6. 64 *bu.*	24. 76.	*Page 86.*	6. $9\frac{?}{?}$.
7. $2.	25. 55.	1. $\frac{71}{8}$.	7. $5\frac{9}{157}$.
8. 6 *firkins.*	26. 83.	2. $\frac{88}{9}$.	8. $6\frac{?}{105}$.
9. 58 *boxes.*		3. $\frac{7.4}{7}$.	9. $9\frac{?}{?}$.
10. 27 *lbs.*	**Art. 59.**	4. $\frac{207}{7}$.	10. $20\frac{?}{?}$.
	Page 78.	5. $\frac{347}{?}$.	11. $3\frac{9}{37}$.
	1. 24.	6. $\frac{114}{?}$.	12. 25.
	2. 2,520.	7. $\frac{710}{?}$.	13. $3\frac{114}{?}$.
		8. $\frac{822}{?}$.	

14. 4.
15. 22⅖.
16. 5⅞.
17. 3.
18. 3.
19. 33.
20. 393.
21. 108.
22. 181.
23. 16.
24. 225.
25. 145.
26. 281.
27. 434.
28. 4.
29. 8.
30. 90.

Art. 70.
Page 89.

1.
2.
3.
4.
5.
6.
7.
8.
9.
10.
11.
12.
13.
14.
15.
16.
17.
18.
19.
20.
21.
22.
23.
24.
25.
26.
27.
28.
29.
30.
31.

32.
33.
34.
55.
36.
37.
38.
39.

Art. 71.
Page 90.

2.
3.
4.
5.
6.
7.
8.
9.

Art. 72.
Page 91.

1.
2.
3.
4.
5.
6.
7.
8.
9.
10.

Art. 73.
Page 92.

3.
4.
5.
6.
7.
8.
9.
10.
11.
12.
13.
14.
15.
16.

17.
18.
19.
20.
21.
22.
23.
24.
25.
26.
27.
28.
29.
30.

Art. 75.
Page 94.

2. 2.
3. 1.
4. 1.
5. 1.
5. 1.
7. 2.
8. 1.
9. 15.
10. 10.
11. 2.
12. 2.
13. 2.
14. 2.
15. 1.
19. 22.
20. 672.
21. 624.
22. 106.
23. 1,357.
24. 134.
25. 11.
26. 20.
27. 15.
28. 13.
29. 8.
30. 14.
31. 23.
32. 26.
33. 3.

34. $24\frac{11}{18}$.
35. $26\frac{28}{36}$.
36. $9\frac{29}{102}$.
37. $10\frac{13}{220}$.
38. $9\frac{103}{850}$.
39. $19\frac{62}{175}$.
40. $10\frac{31}{195}$.
41. $\$6\frac{23}{270}$.
42. $67\frac{31}{364}$ lbs.
43. $486\frac{15}{28}$.
44. $225\frac{87}{106}$ yds.

Problems.
Page 95.

3. $89\frac{3}{4}$ mi.
4. $57\frac{29}{30}$ hrs.
5. $\$394.64\frac{2}{3}$.
6. $\$43\frac{3}{4}$.
7. $123\frac{263}{360}$ lbs.
8. $544\frac{61}{720}$ tons.
9. $135\frac{77}{360}$ yds.
10. $169\frac{1}{4}$ A.

Art. 77.
Page 97.

1. $\frac{1}{36}$.
2. $\frac{1}{12}$.
3. $4\frac{7}{120}$.
4. $1\frac{7}{8}$.
5. $4\frac{1}{12}$.
6. $3\frac{1}{16}$.
7. $11\frac{7}{12}$.
8. $65\frac{7}{22}$.
9. $31\frac{11}{24}$.
10. $\frac{2}{15}$.
11. $52\frac{3}{4}$.
12. $84\frac{13}{20}$.
13. $8\frac{1}{16}$.
14. $2\frac{11}{15}$.
15. $5\frac{7}{12}$.
16. $749\frac{89}{124}$.
17. $\$5\frac{1}{18}$.
18. $6\frac{54}{77}$ ft.
19. $5\frac{149}{600}$.
20. $11\frac{2}{3}$ yds.
21. $20\frac{22}{47}$ yds.
22. $\$94\frac{1}{18}$.
23. $\$12\frac{37}{190}$.
24. $51\frac{757}{952}$.
25. $582\frac{161}{217}$.
26. $110\frac{5}{8}$ bu.

27. $54\frac{31}{40}$ mi.
28. $\$215\frac{11}{16}$.
29. $48\frac{7}{8}$ T.
30. $79\frac{3}{8}$ A.
31. $82\frac{82}{99}$ in.
32. $80\frac{29}{80}$ rds.
33. $\$124\frac{39}{40}$.
34. $201\frac{7}{8}$ rds.
35. $454\frac{7}{8}$ bu.

Problems.
Pages 98–99.

4. $20\frac{39}{50}$ yds.
5. $94\frac{19}{24}$ A.
6. $42\frac{23}{25}$ gals.
7. $94\frac{67}{72}$ tons.
8. $24\frac{1}{2}$ yds.
9. $\$39\frac{1}{10}$.
10. $\$2\frac{3}{8}$.
11. $23\frac{1}{2}$ yds.
12. $33\frac{1}{2}$ mi.
13. $16\frac{1}{2}$ gal.
14. 54 lbs.
15. $\$11\frac{17}{24}$.
16. gain $\$1\frac{3}{25}$.
17. $9\frac{7}{8}$ mi.
18. $12\frac{179}{240}$ yds.
19. lost $\$131\frac{9}{10}$.

Art. 79.
Pages 101, 102.

6. $\frac{7}{40}$.
7. $7\frac{13}{14}$.
8. $9\frac{23}{27}$.
9. $65\frac{5}{8}$.
10. $451\frac{11}{16}$.
11. $2\frac{4}{5}$.
12. 66.
13. $70\frac{5}{12}$.
14. $\frac{19}{64}$.
15. $121\frac{1}{2}$.
16. $1,648\frac{1}{4}$.
17. $10\frac{2}{27}$.
18. $\frac{14}{55}$.
19. $35\frac{7}{12}$.
20. $\frac{3}{5}$.
21. $2\frac{3}{11}$.
22. $21\frac{3}{32}$.
23. $13\frac{3}{9}$.
24. 6.
25. $4\frac{14}{355}$.

26. $9,333\frac{153}{430}$.
27. $\frac{255}{1276}$.
28. $1\frac{1}{4}$.
29. $1\frac{19}{61}$.
30. $1\frac{56}{675}$.
31. $24,450\frac{2}{7}$.
32. $\frac{2}{17}$.
33. $561\frac{33}{64}$.
34. $1,841\frac{7}{64}$.
35. $31,790\frac{41}{56}$.
36. $8,347\frac{15}{16}$.
37. $6,199\frac{43}{64}$.
38. $56,455\frac{55}{63}$.
39. $99,151\frac{4}{5}$.
40. $19,166\frac{83}{110}$.

Problems.
Pages 102, 103.

5. $49\frac{1}{13}$ mi.
7. $13\frac{33}{40}$ bu.
8. $\$43\frac{3}{8}$.
9. $\$17,111\frac{121}{189}$.
10. gain $\$12$.
11. $19\frac{1}{4}$ mi.
12. $\$9\frac{9}{7}$.
13. $322\frac{1}{4}$ A.
14. $28\frac{1}{4}$ mi.
15. $66\frac{1}{2}$ yrs.
16. $\$\frac{3}{8}$.
17. $297\frac{7}{8}$ yds.
18. $185\frac{5}{8}$ mi.

Art. 81.
Pages 105, 106.

3. $\frac{3}{4}$.
4. $\frac{5}{553}$.
5. $1\frac{3}{5}$.
6. $1\frac{3}{8}$.
7. $\frac{188}{189}$.
8. $76\frac{15}{16}$.
9. 1,683.
10. $\frac{110}{189}$.
11. $29\frac{5}{7}$.
12. $2\frac{4}{15}$.
13. $1\frac{1}{4}$.
14. $10\frac{80}{145}$.
15. $1\frac{5}{8}$.
16. $5\frac{54}{115}$.
17. $\frac{7}{810}$.
18. 16.
19. $29\frac{115}{227}$.

20. $1\frac{1027}{2000}$.
21. $1\frac{1}{50}$.
22. $22\frac{1}{4}$.
23. $1\frac{83}{185}$.
24. $5\frac{3}{7}$.
25. $\frac{1}{8}$.
26. $8\frac{53}{98}$.
27. $\frac{77}{440}$.
28. $\frac{23}{38}$.
29. $2\frac{3}{185}$.
30. $1\frac{46}{165}$.
31. $1\frac{97}{350}$.
32. $24\frac{97}{111}$.
33. $\frac{20}{763}$.
34. $\frac{18}{77}$.
35. $1\frac{11}{304}$.
36. $144\frac{5}{8}$.
37. $2\frac{143}{155}$.
38. $6\frac{91}{93}$.
39. $9\frac{7}{18}$.
40. $\frac{14}{285}$.

Problems.
Pages 107, 108.

7. $10\frac{8}{11}$.
8. $5\frac{11}{14}$.
9. $\frac{3}{8}$.
10. $\$104$.
11. $72\frac{8}{11}$ lbs.; $115\frac{85}{111}$ lbs.;
12. $419\frac{11}{14}$ lbs.; $131\frac{3}{4}$ lbs.
13. $1\frac{1}{8}$ da.
14. $\$1\frac{39}{350}$.
15. $29\frac{1}{2}$ lbs.
16. 45 da.
17. $7\frac{1}{4}$ mi.; 42 mi.
18. $\$49,200$.
19. $1\frac{1}{4}$ da.
20. $6\frac{1}{14}$ da.
21. $\$2\frac{1}{2}$.
22. 12 mi.
23. $\$22\frac{79}{144}$.
24. 972 men.
25. $\$20,000$.

Art. 82.
Page 109.

3. 9,850.
4. 93,100.

11

5. 203,075.
6. 255,025.
7. 109,650.
8. 394.
9. 3,724.
10. 3,501.
11. 1,854.
12. 6,961.
13. 1,012½.
14. 11,425.
15. $60,425.
16. $12,500.
17. 948,400 ft.
18. 234,500 yds.
19. $545,625.
20. 426,437½ yds.
21. 810.
22. 914.
23. 730 yds.
24. 405.
25. $873.
26. 261.
27. $18.
28. 80 ft.

Art. 89.
Pages 115, 116.
1. .75.
2. .9375.
3. .421875.
4. .952.
5. .7616.
6. .27168.
7. .508.
8. .048.
9. .0024.
10. 19.875.
11. 24.52.
12. 11.1171875.
13. 110.032.
14. 21.00224.
15. 4.093125.

Art. 90.
Page 117.
1. .1071.
2. .5135.
3. .7143.
4. .1905.
5. 3.8235.
6. 4.9091.

7. 1.625.
8. 1.9636.
9. 6.875.
10. 35.625.
11. 21.6563.
12. 1.0978.
13. 14.9333.
14. 13.094.
15. 4.7848.

Art. 92.
Pages 118, 119.
6. 115.652.
7. 444.0924.
8. $256.017.
9. $144.87.
10. 828.318 lbs.
11. 757.4994 yds.
12. 578.1023.
13. 247.0709.
14. 431.6186 ft.
15. 312.5119.
16. $260.889.
17. 70.1779.
18. 100.001 bu.
19. 122.
20. 81.1027.
21. 155.3006 ft.
22. 684.2371.
23. $3,590.21.
24. 231.3898 bu.
25. 48.917 yds.

Problems.
Pages 119, 120.
2. 37.495 A.
3. $83.92.
4. $47.39.
5. 149.731 yds.
6. 330.275 bu.
7. 44.979 C.
8. $56,186.97½.

Art. 94.
Page 122.
12. 71.507.
13. 66.9997.
14. 887.8002.
15. $9.77.
16. 1.8705 ft.
17. $565.928.
18. $24,998.923.

19. .999 yds.
20. $.999.
21. 9.9997 ft.
22. 5.9994 yds.
23. $4.128.
24. 4.2197.
25. 6.5766.
26. 4.3219.
27. 4.8986.
28. $25.453.
29. 76.112 lbs.
30. $11.075.
31. 127.61 lbs.
32. $2,761.985.
33. $333.13.

Problems.
Pages 122, 123.
1. $4.05½.
2. 25.48 yds.
3. $170.133.
4. $305.37½.
5. $3,107.25.
6. $939.49.
7. $2,045.35.
8. $783.80.
9. $963.68.
10. $4,727.57.
11. $38.29½.
12. 15.684 C.
13. 74.39 ft.
14. $57.05.
15. 123.315 lbs.
16. 199.85 mi.
17. $82.75.
18. gain $1.16½.

NOTE.— Answers
are carried to four
places in Articles
96-98.

Art. 96.
Pages 125, 126.
3. $641.28.
4. $5 3438.
5. 0.21 bu.
6. $0.0012.
7. 18,782.03 yds.
8. 0.0904 ft.
9. 0.0004 lbs.
10. 42.102.3603.
11. 1,744.3913.

12. 23.8894.
13. 386.5576.
14. $8,361.32.
15. $3.0088.
16. $0.2475.
17. $30.1534.
18. $63.8406.
19. $0.0027.
20. 20.4905 lbs.
21. 84.5688.
22. 0.3737.
23. 412.5508.
24. 0.1197 yds.
25. 0.5103 rds.
26. 87.4894.
27. $1,713.782.
28. 964.5215.
29. 72.5641.
30. 29,170.4499.
31. 3.3538.
32. $1,485.
33. 81.648 lbs.
34. 1,594.974 yds.
38. $117.33.
39. 9,189.866 lbs.
40. $4,812.975.
41. 533.844 yds.
42. 3,512.7728.

Problems.
Pages 126-128.
1. $108.37½.
2. $117.56½.
3. $200.59½.
4. $129.68¾.
5. $2,675.75.
6. $12.75.
7. $27.18¾.
8. $14,114.25.
9. $41.78⅛.
10. $5.699.
11. $23.751.
12. $192.231.
13. $81.56¼.
14. 32.045 mi.
15. 264.1875 mi.
16. 65.515 mi.
17. $13.737.
18. $111.58.
19 $14.396 gain.
20. 297.2538 mi.

Art. 98.

Pages 130, 131.

4. 0.0025.
5. $1.39.
6. 21.5434.
7. 0.25.
8. 87.5.
9. 4.75.
10. 112.8767.
11. 356.1111.
12. 12.5 *ft.*
13. 12.24.
14. 1,485.6016.
15. $0.167.
16. 153.8462.
17. $0.06¼.
18. 32.27 *yds.*
19. 0.0268.
20. 0.916.
21. $0.67.
22. 2.36.
23. 79.52.
24. 123.107.
25. 7.54.
26. $70.55.
30. $16.196.
31. $1.439.
32. 7.1364 *ft.*
33. 34.775 *yds.*
34. 3.1034 *lbs.*

Problems.

Pages 131–133.

5. 48.5 *A.*
6. $8.50; $53.55.
7. 19 *C.*; 3.7 *C.*
8. 17.5 *yds.*; 35.6365 *yds.*
9. 149 *lbs.*
10. 87.5 *bbls.*
11. 13.5 *mi.*; 15.25 *mi.*
12. $3.
13. 58.8666 *mi.*; 559.2333 *mi.*
14. 306.675 *mi.*
15. 10.4 *hrs.*
16. $128.50.
17. $1.12½; $7.31¼.
18. 16.08\frac{9}{10}$.
19. $67.68.
20. 29⅓ *cts.*
21. 11 *hrs.*
22. 72 *bu.*
23. 18 *cts.*; $5.62½.
24. 51 *bu.*
25. 23⅓ *bu.*
26. 215.18 *mi.*
27. 10.5.
28. 7 *hrs.*
29. $27.50.
30. 10.5 *C.*
31. $5.89.
32. 41 *T.*
33. 7 *vests.*

Art. 100.

Pages 134, 135.

3. $62.50.
4. $18.
5. $3.75.
6. $23.
7. $3.12½.
8. $9.12½.
9. $4.25.
10. $23.40.
11. $15.66⅔.
12. $9.06¼.
13. $29.10.
14. $39.37½.
15. $23.33⅓.
16. $7.

Art. 101.

Pages 135, 136.

2. 5.17\frac{8}{10}$.
3. $356.66¼.
4. $54.46¼.
5. $5,025.
6. $34.33½.
7. $35.77.
8. $118.14.
9. $262.23.
10. $82.87½.
11. $219.60.
12. $37.80.
13. $69.
14. $70.61.

Art. 102.

Page 136.

2. $23 59$\frac{7}{10}$.
3. 99.53\frac{7}{10}$.
4. $79.67½.
5. $407.10.
6. $1,029.60.
7. $12.96.
8. $63.602.
9. $57.838.
10. $66.18.
11. $70.18.
12. $8.32½.
13. $22.06¼.
14. $301.883.

Art. 103.

Pages 138, 139.

1. 102.58\frac{4}{10}$.
2. $111.83.
3. $51.
4. $6.62.
5. $89.87.
6. $49.81.

Art. 104.

Page 140.

1. $86.28 *Cr.*
2. $247.68 *Cr.*

Art. 146.

Pages 159, 160.

1. 4,382 *far.*
2. 21,268 *grs.*
3. 37,740 *min.*
4. 5,436 *yds.*
5. 10,890 *sq. ft.*
6. 565 *pts.*
7. 2,739,600 *sec.*
8. 30,183 *far.*
9. 1,365 *in.*
10. 175,000 *sq. li.*
11. 200 *cu. ft.*; 3,200 *cu. ft.*
12. 664 *qts.*
15. 17*s.* 6*d.*
16. 11 *oz.* 5 *dwts.*
17. 2 *cwt.* 1 *qr.* 12 *lbs* 8 *oz.*
18. 4 *sq. ft.* 7.2 *sq. in.*
19. £2 7*s.* 6*d.*
20. 3 *da.* 12 *hrs.* 15 *min.*
21. 2 *yrs.* 37 *w.* 5 *da.* 6 *hrs.*

22. 4 *T.* 11 *cwt.*
23. 3 *T.* 16 *cwt.* $12\frac{7}{8}$ *lbs.*
24. 2 *mi.* 6 *fur.* 30 *rds.*
25. 7 *fur.* 21 *rds.* $5\frac{9}{11}$ *in.*
26. 3 *bu.* 2 *pks.* 3 *qts.*
27, 43 *yds.* $6\frac{3}{8}$ *in.*
28. 17 *yds.* 1 *ft.* $1\frac{1}{2}$ *in.*
29. 4 *hhds.* 1 *bbl.* 3 *gals.* 3 *qts.* 1.248 *pts.*
30. 3° 42′.
31. 12° 9′ 43″.2.
32. 8 *cwt.* 2 *qrs.* 22 *lbs.*
33. 17*s.* $0\frac{4}{9}$*d.*
34. 4 *bu.* 1 *pk.* $3\frac{15}{19}$ *qts.*
35. 2 *T.* 11 *cwt.* 20 *lbs.*
36. 3 *da.* 13 *hrs.* 36 *min.* 28.8 *sec.*
37. 4 *fur.* 23 *rds.* 3 *yds.* 2 *ft.* 2.64 *in.*
38. $89\frac{1}{2}$ *cts.*
39. 3 *A.* 1 *R.* 33.6 *sq. rds.*
40. 1 *qt.* 1.656 *pts.*
41. 3 *fur.* 8 *rds.* 3 *yds.*

Art. 148.

Pages 162–164.

1. 2 *lbs.* 8 *oz.* 15 *dwts.* 12 *grs.*
2. 52 *w.* 1 *da.* 6 *hrs.*
3. 8℔ 6℥ 3ʒ 2℈.
4. 7℥ 1℈ 12 *grs.*
6. 6 *rds.* 4 *yds.* 2 *ft.* 9 *in.*
7. 4 *mi.* 97 *rds.* 1 *ft.* 10 *in.*
8. 1 *mi.* 126 *rds.* 3 *yds.* 1 *ft.* 7 *in.*
9. 2 *qrs.* 4 *lbs.* 9 *oz.*
10. 238 *gals.* 2 *qts.* 2 *gi.*
11. £8 14*s.* 5*d.*
12. 2 *lbs.* 5 *oz.* 16 *dwts.* 7 *gr.*
13. 16℔ 6℥ 6ʒ 1℈.
14. 10 *cwt.* 1 *qr.* 11 *oz.*
15. 3 *da.* 14 *hrs.* 29 *m.* 35 *sec.*
16. 3 *wks.* 20 *hrs.* 23 *m.*
17. 5° 51′ 58″.
18. 18° 34′.
19. 4.643 *m.*
20. 10 *sq. yds.* 2 *sq. ft.* 114 *sq. in.*
21. 12 *A.* 1 *sq. ch.* 1,312 *sq. li.*
23. $\frac{7}{8}$ *T.*
24. $\frac{1}{18}$ *hhd.*
25. $\frac{4}{11}$ *mi.*
26. $\frac{122}{3375}$ *rt. a.*
27. $\frac{21}{32}$ *gal.*
28. $\frac{71}{128}$ *C.*
29. 0.714 *m.*
32. £14.8625.
33. 7.871 *mi.*
34. 0.1188 *T.*
35. 0.2709℔.
36. 0.2917 *gal.*
37. 0.5833 *yd.*
38. 28.25 *oz.*
39. 4,475 *lbs.*
40. 35.75 *pks.*
41. 104.75*s.*
42. 109.375 *gals.*
43. 263.1667 *yds.*
44. 2,855′.7.
45. 0.043 *decam.*
47. $\frac{1010}{3263}$.
48. $\frac{11}{19}$.
49. $\frac{71}{462}$.
50. $\frac{1}{6}$.
51. $\frac{171}{8960}$.
52. $\frac{3}{5}$.

Miscellaneous.

Pages 165, 166.

1. 11,400 *grs.*
2. 16 *lbs.* 10 *oz.* 18 *dwts.* 5 *grs.*
3. 49,775 *lbs.*
4. 4,368 *qts.*
5. 24.8 *mi.*
6. $20\frac{1}{3}$ *C.*
7. 214 *da.* 15 *hrs.* 30 *m.* 35 *sec.*; 30 *wks.* 4 *da.* 15 *hrs.* 30 *m.* 35 *sec.*
8. $4\frac{1}{3}$*d.*; 4*s.* 8*d.*
9. 275.59*in.*; 511.81*in.*
10. 15.367 *m.*; 8.1072 *m.*
11. 7.4564 *mi.*
12. 24.1402 *kilom.*; 12.0701 *kilom.*
13. 43.3247 *qts.*
14. 105.668.
15. 119.2391 *l.*; 280.1173 *l.*
16. 25.077 *g.*: 1.7988 *g.*
17. 163.1404 *lbs.*
18. 1,124.346 *lbs.*
19. 4.1385 *C.*
20. 4.8283 *C.*; 618.0224 *cu. ft.*
21. £0.8906.
22. $\frac{23}{270}$.
23. 1.7526 *lbs.*; 21.0312 *oz.*
24. 0.5144 *wks.*
25. $\frac{17}{63}$.
26. $\frac{108}{440}$.
27. $\frac{15}{32}$.
28. $\frac{28}{80}$.
29. 3*s.* 8*d.* 3.52 *far.*
30. 4 *fur.* 23 *rds.* 11 *ft.* 2.64 *in.*

Art. 150.

Pages 168–170.

4. 83 *bu.* 6 *qts.*
5. 2 *bu.* 2 *pks.* $2\frac{1}{2}$ *qts.*
6. 27 *gals.* 2 *qts.* 1 *pt.*
7. 16 *yds.* 2 *ft.* 5 *in.*
8. 19 *A.* 3 *sq. ch.* 1926 *sq. li.*
9. 19 *da.* 23 *hrs.* 28 *min.*
10. £35 15*s.*
11. 107*lbs.* 4*oz.* 10*dwts.*
12. 12 *mi.* 306 *rds.*
13. £92 1*d.*
14. 157 *bu.* 4 *qts.*
16. £1.8.
17. 5.0987 *da.*
18. £49 2*s.* $2\frac{1}{2}$*d.*
19. 2 *lbs.* 6 *oz.* 19 *dwts.* 17 *grs.*
20. 1℔ 3℥ 3ʒ 2℈ 18 *grs.*
21. 1 *T.* 10 *cwt.* 2 *qrs* 4 *lbs.*
22. 39 *wks.* 15 *hrs.*
23. 25 *mi.* 253 *rds.* 1 *yd.* 1 *ft.* 6 *in.*

24. $59\frac{1}{8}$ *yds.*
25. 221 *bu.* 2 *pks.* 5 *qts.*
26. 34.022 *T.*
27. 180.5357 *lbs.*
28. £35.2375.
29. 89.1 *da.*

Problems.
Pages 170, 171.

1. 103 *cwt.* 10 *lbs.*
2. 21 *lbs.* 4 *oz.*
3. $297\frac{7}{8}$ *yds.*
4. 293 *A.* 6 *sq. rds.*
5. 22 *yrs.* 5 *mo.*
6. $22.718.
7. 91.094 *lbs.*
8. 67 *A.* 7 *sq. ch.*
9. $134\frac{15}{16}$ *yds.*
10. 4 *C.* 36 *cu. ft.*

Art. 152.
Pages 173–175.

4. 2 *cwt.* 1 *qr.* 23 *lbs.*
5. 36 *hhds.* 49 *gals.*
6. 2 *yds.* 2 *in.*
7. 4 *A.* 145 *sq. rds.*
8. 15° 55′ 57″.
9. 17℔ 8ʒ 4ʒ 2Ɔ 16 *grs.*
11. 8 *rds.* 4 *yds.* 9 *in.*
12. 4 *rds.* 3 *yds.* 1 *ft.*
14. 4 *yrs.* 9 *mos.* 9 *da.*
15. 58 *yrs.* 3 *mos.* 24 *da.*
16. 43 *yrs.* 1 *mo.* 12 *da.*
17. 8 *hrs.* 7 *min.* 13 *sec.*
18. £1 2*s.* 6*d.*
19. 16 *gals.* 2 *qts.* $2\frac{36}{55}$ *gi.*
20. 2,775.8 *grams.*
21. 3 *da.* 22 *hrs.* 43 *m.* 34 *sec.*
22. 3 *T.* 16 *cwt.* 2 *qrs.* 15 *lbs.*
23. 5 *lbs.* 7 *dwts.* 13 *grs.*
24. 684 *gals.* 2 *qts.* 1 *pt.*
25. 15 *bu.* 1 *pk.* 3 *qts.*

Problems.
Pages 175, 176.

2. 161 *A.* 87 *sq. rds.*

3. $31\frac{18}{19}$ *yds.*
4. 104 *cwt.* 1 *qr.* 10 *lbs.*
5. 2 *yrs.* 9 *mo.* 24 *da.*
6. 72 *yrs.* 8 *mo.* 11 *da.*
7. 7 *yrs.* 9 *mo.* 1 *da.*
8. 283 *yrs.* 8 *mo.* 23 *da.*
9. 6 *C.* 6 *C. ft.* 4 *cu. ft.*
12. 6° 18′ 30″.
14. 90° 24′ 15″.
15. 8° 52′ 8″.
16. 37 *bu.* 2 *pks.*

Art. 154.
Pages 178–180.

4. 1 *T.* 18 *cwt.* 2 *qrs.*
5. $214.37\frac{1}{2}$.
6. 3,006.1 *fr.*
7. 70.08 *kilog.*
8. 322 *yds.*
9. 1848.75 *m.*
10. 83 *wks.* 4 *da.*
11. 5 *hhds.*
12. 2 *cwt.* 1 *qr.* 18 *lbs.*
13. 124 *yds.*
14. £11 13*s.* 4*d.*
15. 112 *A.* 3 *R.*
16. 3 *da.* 6 *hrs.* 4 *min.*
17. 6 *T.* 14 *cwt.* 1 *qr.* 15 *lbs.*
18. £123 9*s.* 6*d.*
19. 2,317 *T.* 15 *cwt.*
20. 87 *qrs.* 1 *ft.* 6 *in.*
21. $114\frac{2}{25}$ *gals.*
25. 9,656 *sq. yds.*
26. $308\frac{1}{8}$ *sq. rds.*
27. 747.1111 *cu. yds.*
29. 28.632 *cu. ft.*
30. 1,017.1875 *cu. ft.*; 131.25 *sq. ft.*
31. 257.375 *sq. ft.*

Problems.
Pages 180, 181.

1. £501 17*s.* 6*d.*
2. 959 *mi.* 20 *yds.*
3. 452 *lbs.* 6 *oz.* 4 *dwts.* 6 *grs.*
4. 6210 *yds.*
5. 16 *C.* 5 *C. v.*; 2128 *cu. ft.*

6. 38 *oz.* 14 *dwts.*
7. 1 *T.* 17 *cwt.* 2 *qrs.* 23 *lbs.*
8. $37\frac{1}{2}$ *mi.*
9. 78 *A.* 30 *sq. rds.*
10. £41 3*s.* 6*d.*
11. 237 *yds.*
12. $183.31\frac{1}{4}$.
13. 255.125 *yds.*
14. 17 *T.* 5 *cwts.*
15. 1,227.25 *sq. yds.*
16. 444.375 *sq. ft.*
17. 27.236 *yds.*; 245.125 *sq. ft.*; 67.9722 *sq. yds.*
18. 7,020 *lbs.*

Art. 156.
Pages 184, 185.

3. 2 *cwt.* 1 *qr.* 18 *lbs.* $3\frac{1}{4}$ *oz.*
4. 20 *T.* 16 *cwt.* 1 *qr.* $1\frac{1}{4}$ *lbs.*
6. 2 *mi.* 4 *fur.* 1 *rd.*
7. 2 *bu.* 3 *pks.* 4 *qts.*
8. £1 8*s.* 9*d.*
9. 1 *lb.* 3 *oz.* 6 *dwts.*
10. 5.2 *kilog.*
11. 704.
12. 48 *mi.* 201 *rds.*
14. .09 *fr.*
14. 35 *liters.*
15. £4 6*s.* $8\frac{8}{9}d.$
16. 49 *gals.* $2\frac{1}{2}$ *qts.*
17. 7,040.
18. 1.42 *fr.*
19. .42 *kilog.*
20. 16.
21. 17.
22. 3.783.
23. 5° 13′ 5″.
24. 4° 36′ 19″.
25. 3.667.
26. 400.
27. 2,880.
28. 4.5.
29. 15.
30. 1.0833.
32. £1 11*s.* 5*d.*
33. 2 *lbs.* 3 *oz.* 17 *dwts.* 8 *grs.*

34. 29.575.
35. 377.

Problems.
Pages 185–188.

1. £2 3s.; £10 15s.; £27 19s.
2. 10¼ mi.; 31¼ mi.
3. 35 *times*.
4. 24 *yds*.
5. 28½ *yds*.
6. 9 *in*.
7. $990.
8. 80 *panes*.
9. $7.98.
10. $73.50.
11. $24.36.
12. $20.80.
13. $130.
14. 720.
15. 22.
16. 2 A. 1 R. 30 *sq.yds*.
17. 12 *lots*.
18. $147.
19. $1,367½.
20. 8,400 *bricks*.
21. 768.
22. $34.222.
23. $82.50.
24. 54.
25. 1,920.
26. 9s. 5¼d.
27. 1 C. 22.86 *cu. ft*.
28. £1 12s. 2d.; £19 6s.
29. 84.25 *fr*.
30. 6 *hrs*. 48 *min*.
31. 882.65 *fr*.
32. 7 *mi*. 1180 *yds*.; 9 *mi*. 1260 *yds*.
33. 270 *yds*.
34. 47° 15′ 16″.
35. 86° 49′ 26″.

Art. 157.
Page 189.

2. 1 *hr*. 9 *m*. 37 *sec*.
3. 3 *hrs*. 37 *m*. 15 *sec*.
4. 7 *hrs*. 53 *m*. 34 *sec*.
5. 1 *hr*. 27 *m*. 11 *sec*.
6. 5 *hrs*. 18 *m*. 41 *sec*.

7. 2 *hrs*. 34 *m*. 35 *sec*.
8. 46° 1′ 30″.
9. 32° 19′ 30″.
10. 78° 35′ 45″.
11. 139° 17′ 30″.
12. 19° 39′.
13. 97° 16′.
14. 2*hrs*. 37*min*. 29.467 *sec*.
15. 70° 57′.

Art. 165.
Page 193.

2. 63.99 *lbs*.
3. $39.
4. 56.7 *yds*.
5. 13.23 *mi*.
5. $30.
7. 1 *ct*.
8. 3.4375 *kilog*.
9. 122½ *ft*.
10. 5.1985.
11. $18.
12. 204.54 *kilog*.
13. 4.2636 *lbs*.
14. $115.981.
15. $21.058.
16. £.0925.
17. 15¾ *yds*.
19. 200 A.
20. £21 12s. 6d.

Art. 166.
Page 194.

5. $218; $182.
6. 962½ *lbs*.; 577½ *lbs*.
7. 64 *yds*.; 30 *yds*.
8. 45.36 *kilom*.; 38.64 *kilom*.
9. 125 *hhds*.; 25 *hhds*.
10. 40.32 *yrs*.; 31.68 *yrs*.
11. 163.4 *ft*.; 140.6 *ft*.
12. 74.37 *mi*.; 59.63 *mi*.
13. 86.72 *T*.; 41.28 *T*.
14. $99.76.; $72.24.
15. 78.48 *bu*.; 65.52 *bu*.
16. 66.08 *fr*.; 45.92 *fr*.
17. $92.
18. 53.1 A.

19. $5.62½.
20. 8½ *cts*.

Art. 167.
Pages 195, 196.

4. 6%.
5. 43¾%.
6. 1⅜%.
7. 13½%.
8. 10¹⁴⁄₁₅%.
9. 460⁴⁄₃₃%.
10. 124%.
11. 166⅔%.
12. 20%.
13. 70%.
14. 25%.
15. 74.2%.
16. 87½%.
17. 34%.
18. 60%.

Art. 168.
Pages 196, 197.

4. $1,500.
5. $420.
6. 15,465.
7. 66⅔ *gals*.
8. $4,500.
9. 180.
10. 50 *rds*.
11. 60 *ft*.
12. 200 *sheep*.

Problems.
Pages 197–199.

1. $24.25.
2. 1,304.75.
5. £1,856 17s. 6d.
4. $4.
5. 550.
6. 690 *eggs*.
7. $50,000.
8. 13,000.
9. 4.8 *kilom*.
10. 10.67 *hectol*.
11. $5,400.
12. $740.
13. 53 *gals*.
14. $20,000.
15. 31,000 *men*.

16. $3,026.98½.
17. $2,540.62½.
18. 676.
19. 234⅗ yds.
20. 12½%.
21. $3,312½.
22. 47½%.

Art. 173.

Pages 200–201.

2. $168.
3. $23,482.81¼.
4. $4,080.
5. $52.50.
6. $300.93¾.
7. 2%.
8. $60.37½.
9. $585.75.
11. 1,000.
12. 140,000 *lbs.*
13. $43.923.
14. $47.25.
15. $150 15/16; $11,924 1/16.
16. 42,997½ *lbs.*
17. $3,145.60.
18. $262.50.

Art. 177.

Pages 204–205.

2. $2400¼.
3. $66.25.
4. $277.50.
5. $468.75.
6. $1,137.
7. $21.
8. $90.75.
9. $17,040.
10. $14,000.
13. $69.
14. $232.50.
15. $875.
16. $618.784.
17. $2,223.75.
20. $713.81¼.
21. $60.

Art. 179.

Pages 206–209.

2. $1,050.
3. $58.50.
5. $20,160.
6. $24.
7. £375.
8. $2,500.
9. $35,112.
10. $550.
11. 12 *cts.*
12. $50,000.
13. $9,000.
14. $325.
15. $330.
16. $23.88.
17. 50%.
18. $12,400.
19. $3,447.50.
20. $5,636.25.
21. 25%.
22. $1,000.
23. $120.
24. $135.
25. 36 4/11%.
26. 36 *cts.*
27. $27,500.
28. $160.
29. $7.50.
30. 15%.
31. 16%.
32. $282.
33. 22½%.
34. 25%.
35. 30%, or $4.20.

Art. 182.

Page 210.

3. $78.61.
4. $25.

Art. 186.

Page 213.

2. $194.53.
3. $2,470.228.
4. $915.60.
5. $128.16.
6. $46.13.
7. $64.11.
8. $554.40.
9. $45.10.
16. $432.64.
11. $210.
12. $158.60.
14. $4,416.494.
15. £179 15*s.* 2*d.* 1⅗ *far.*
17. $517.50.
18. $4,795.56.
19. £3,166 13*s.* 4*d.*
20. $8,718.75.
21. $1,030.40.
22. $2,115.80.
23. $3,392.50.
24. $11,732.58.
25. $18,597.96.
26. $1,080.
27. $1,200.

Art. 187.

Pages 214–217.

3. $58.892.
4. $28.35.
5. $66.36.
6. $1,387.80.
8. $4,639.70.
9. $2,955.767.
10. $18.23.
11. $2.65.
12. $729.646.
14. $44.50.
15. $82.25.
16. $393.42⅜.
17. $63.45.
18. $54.37½.
19. $918.60.
20. $1,012.65.
23. $510.86.
24. $191.205.
25. $266.07.
26. $328.32.
27. $190.151.
28. $44.289.
29. $7.66⅔.
30. $55.417.
31. $193.61.
32. £52 10*s.*
33. $79.875.
34. $190.517.
37. £12,170.
38. $372.60.
39. $4,606.87½.
40. $6,974.57½.
41. $843.
42. $297.50.
43. $3,875.37½.
44. $964.219.
45. $328.
46. $74.025.
47. $71.867.
48. $532.50.
49. $4,205.33⅓.
50. $5,385.

Art. 188.

Pages 217, 218.

2. $0.526.
3. $144.60.
4. $1.302.
5. $0,72½.
6. $1.449.
7. $11.828.
10. $7.92; $13.20.
11. $4.37; $7.29.
12. $1.008; $2.176.
13. $1.408; $4.224.
14. $1.087; $1.521.

Art. 189.

Page 219.

2. $7.84.
3. $20.16.
4. $66.21.
5. $1.167.
6. $4.792.
7. $23.123.

Art. 190.

Page 219.

1. 7%.
2. 4½%.
3. 7%.
4. 6%.
5. 7%.
6. 7%.
7. 5%.

Art. 191.

Page 220.

1. 3 *yrs.*
2. $3\frac{1}{4}$ *yrs.*
3. 2 *yrs.*
4. 9 *mos.*
5. 1 *yr.* 2 *mos.*
6. 1 *yr.* 3 *mos.*
7. $17\frac{1}{2}$ *yrs.*
8. 2 *yrs.*
9. 4 *yrs.*

Art. 192.

Page 221.

1. \$2,100.
2. \$3,000.
3. \$2,750.
4. \$1,800.
5. \$2,500.
6. \$100.

Miscellaneous.

1. \$950.
2. 7%.
3. 5 *yrs.*
4. \$225.
5. 7 *yrs.* 4 *mo.*
6. \$1,500.
7. 2 *yrs.*
8. 14%.
9. \$646.333.
10. 7 *yrs.* 11 *mo.* 24 *da.*

Art. 193.

Page 222.

2. \$371.28.
3. \$351.232.

Art. 198.

Pages 225, 226.

2. \$252.074.
3. \$200.55.
5. \$128.152.
6. \$5,081.628.
7. \$474.887.
8. \$700.452.
10. \$405.44.

Art. 200.

Pages 229, 230.

2. \$175.253.
3. \$823.401.
5. \$1,671.75.
6. \$496.17.

Art. 202.

Pages 231, 232.

2. \$1,055.70.
3. $\$10.62\frac{1}{2}$.
4. \$30.
5. \$803.25.
6. $\$494.49\frac{2}{5}$.
7. $\$346.87\frac{1}{2}$.
8. \$4.95.
9. $\$.59\frac{3}{8}$.

Art. 203.

Pages 233, 234.

3. \$1,454.546.
4. \$41.25.
5. \$66.364.
6. \$850.
7. \$136.
8. \$1,090.909.
9. \$60.287.
10. \$680.581.
11. \$1,095.506.
12. \$9,830.425.
13. \$1,750.
14. \$4,000.
15. \$3,420.089.

Art. 207.

Page 236.

3. \$7.035.
4. \$6.329.
5. \$989.50.
6. \$1,696.752.
7. \$1.593.

Art. 208.

Page 237.

2. \$505.306.
3. \$765.853.
4. \$1,009.251.

Art. 218.

Pages 240, 241.

6. \$2,090.
7. \$105,322.50.
8. \$168.75.
9. $\$112\frac{3}{4}\%$.
10. \$10,125.
11. $111\frac{3}{4}\%$.
12. \$8,000.
15. $5\frac{285}{343}\%$.
16. $7\frac{1}{7}\%$.
18. $71\frac{3}{7}\%$.

Art. 227.

Page 246.

11. $\$1,971.41\frac{1}{2}$.

Art. 229.

Page 247.

2. \$3,928.50.
3. \$1,560.06.
4. \$1,587.60.
6. £563 9*s.* $8\frac{3}{4}d$.
8. \$6,057.143.
9. 5,200 *fr.*

Art. 233.

Page 250.

5. 70 *da.*

Art. 246.

Pages 260, 261.

1. $\frac{5}{3}=1\frac{2}{3}$.
2. 5.
3. 5.
4. 3.
5. $\frac{1}{4}$.
6. 10.
7. 30.
8. $\frac{1}{16}$.
9. $\frac{1}{8}$.
10. $\frac{1}{5}$.
11. 28.
12. $\frac{1}{16}$.
13. 2 *lbs.* 10 *oz.*
14. 3 *lbs.* 1 *oz.*
15. \$3.
16. 20 *yds.*

Art. 250.

Page 263.

3. 20 *yds.*
4. 27.
5. 96.
6. 9.
7. $\frac{1}{6}$.
8. 0.26.
9. \$4.20.
10. $£3\frac{1}{2}$.
11. $\$270.83\frac{1}{3}$.
12. \$7.50.
13. £4 8*s.* 6*d.*
14. $3\frac{1}{4}$ *C.*
15. \$165.79.
16. 49.
17. \$5.
18. $64\frac{1}{2}$.
19. \$1.50.
20. $\$52\frac{1}{2}$.
21. \$20.909.
22. $6\frac{1}{7}$ *yds.*

Art. 252.

Pages 265–268.

3. $x=643\frac{1}{2}$ *mi.*
4. $x=58.75$.
5. $x=\$125.797$.
6. $x=\$190.85$.
7. $x=28$.
8. $x=\$3.29$.
9. $x=11$.
10. $x=300$.
11. $x=86\frac{86}{469}$ *da.*
12. $x=14\frac{1}{7}$ *yds.*
13. $x=48$ *men.*
14. $x=1\frac{1}{4}$ *yds.*
15. $x=1\frac{11}{25}$ *da.*
16. $x=\$3$.
17. $x=20$ *bu.*
18. $x=\$8.22\frac{1}{2}$.
19. $x=16\frac{2}{3}$ *da.*
20. $x=\$9.95$.

21. $x=\$132$.
22. $x=112\frac{1}{2}$ *bu.*
23. $x=33.15$ *fr.*
24. $x=35$ *m.*
25. $x=\$320$.
26. $x=162$ *da.*
27. $x=\$3,597.22\frac{2}{9}$.
28. $x=\$9,972.97\frac{3}{10}$.
29. $x=\$7,200$.
32. $x=10$ *horses.*

Art. 253.

Page 269.

5. A., $180; B., $85; C., $240.
6. A., $37½; B., $62½.
7. A., $130; B., $195; C., $325.
8. A., $150; B., $225; C., $270; D., $375.

Art. 258.

Page 276.

1. 64.
2. 625.
3. 2,744.
4. 625.
5. 9,604.
6. .0081.
7. .0225.
8. 15,625.
9. .0359.
10. 11.56.
11. $\frac{27}{64}$.
12. $\frac{16}{81}$.
13. $15\frac{5}{8}$.
14. $34\frac{221}{224}$.
15. $410\frac{1}{16}$.

Art. 261.

Page 000.

2. 98.
3. 115.
4. 585.
5. 972.
6. 121.
7. 194.
8. 204.
9. 229.
10. $\frac{11}{12}$.
11. $\frac{5}{27}$.
12. $\frac{13}{17}$.
13. $\frac{5}{19}$.

Miscellaneous Examples.

1. 369.
2. 15.
3. 2. 2. 2. 5. 5. 7. 7.
4. 2. 3. 5. 7. 19.
5. 458.
6. 35.
7. 1,800.
8. 3,556.
9. $2\frac{27}{200}$.
10. $1\frac{1}{7}$.
11. $\frac{11}{60}$.
12. $1\frac{7}{18}$.
13. 34.639.
14. 1.2062.
15. $240.
16. 480.
17. A., $51; B., $127½; C., $178½.
18. $1\frac{13}{47}$ *da.*
19. 214 *bu.* 3 *pks.* 4 *qts.*
20. 12 *bu.*
21. $13\frac{19}{20}$ *cts.*
22. $250; $350; $400.
23. $273.
24. 10 *mi.*
25. 47,250 *lbs.*
26. 60 *da.*
27. $4,070.
28. $160.
29. $4,350.
30. $898.80.
31. 18%.
32. 25%.
33. 40%.
34. 53 *mi.*
35. $15.
36. 120 *rds.*
37. 195 *bu.*
38. 30 *men.*
39. $330.
40. 214 *sheep.*
41. 25,920 *men.*
42. 250 *melons.*
43. 216 *ft.*
44. S., $660; J., $577.50; B., $412.50.
45. $\$2,604\frac{1}{6}$.
46. $\$4,102\frac{22}{39}$.
47. 18 *bales.*
48. $2\frac{1}{2}$ *mo.*
49. $36.75.
50. $131.25.
51. 1st, $15; 2d, $22.50; 3d, $7.50.
52. $11,666\frac{2}{3}$ *bd.ft.*
53. 21 *C.* $13\frac{1}{8}$ *cu. ft.*
54. A., $128; B., $168; C., $104.
55. $1 *per yd.*
56. 50 *ft.*
57. $32,000.
58. 25%.
59. 2,400
60. 9 *weeks.*
61. 3,468 *sq. ft.*
62. *Son,* $8,000; *eldest dau.,* $6,000; *younger dau.* $4,000.
63. 286,688.
64. 120 *men.*
65. 14.
66. 5 *hrs.* $27\frac{3}{11}$ *min.*
67. 750 *men.*
68. 66 *yrs.*
69. $28\frac{2}{7}$ *da.*
70. 12 *hrs.* 10 *m.* P.M.
71. 125 *reams.*
72. 12 *da.*
73. 75 *cu. yds.*
74. 11 *da.*
75. 23 *A.*
76. 63 *kilog.*
77. 15.76 *m.*
78. 167.4.
79. 39.06 *kilog.*
80. *Man* 6.15 *fr.*; *wom.* 2.05 *fr.*
81. A., $280; B., $420.
82. W., $4,420. B., $2,080.
83. A., $4,080; B., $2.100; C., $1,920.
84. A., $7,219.84; B., $\$5,086.70\frac{6}{11}$; C., $\$5,743.05\frac{5}{11}$.
85. 5 *better,* 4 *poorer.*
86. $16\frac{4}{5}$ *da.*
87. *Elder,* $5.400; *younger,* $4,600.
88. $1,500.
89. $4,800.
90. 16 *men.*
91. 72 *in.*
92. B., $3,750; C., $3.125.
93. *Capt.,* $243; *man,* $162; *boy,* $54.
94. $\$1,666\frac{2}{3}$.
95. 190.
96. A., $445; B., $230; C., $325.
97. $10\frac{191}{2240}$.
98. 5 *hrs.* 20 *m.*
99. 45 *yrs.*
100. 65.